Construction Economics and Building Design:
A Historical Approach

Construction Economics and Building Design: A Historical Approach

R. Gregory Turner, AIA

Illustrations by R. Gregory Turner, AIA

VNR | VAN NOSTRAND REINHOLD COMPANY
_____ New York

Copyright © 1986 by Van Nostrand Reinhold Company Inc.
Library of Congress Catalog Card Number 85-15759
ISBN 0-442-28309-1

All illustrations were prepared by R. Gregory Turner, AIA.
Some were adapted from other sources, credit for which appears
with the individual drawing.

Printed in the United States of America
Designed by Barbara M. Marks

Published by Van Nostrand Reinhold Company Inc.
115 Fifth Avenue
New York, NY 10003

Van Nostrand Reinhold Company Limited
Molly Millars Lane
Wokingham, Berkshire RG11 2PY, England

Van Nostrand Reinhold
480 La Trobe Street
Melbourne, Victoria 3000, Australia

Macmillan of Canada
Division of Gage Publishing Limited
164 Commander Boulevard
Agincourt, Ontario M1S 3C7, Canada

16 15 14 13 12 11 10 9 8 7 6 5 4 3 2 1

Library of Congress Cataloging-in-Publication Data

Turner, R. Gregory, 1952–
    Construction economics and building design.

    Bibliography: p.
    Includes index.
    1. Building—Estimates.   2. Architecture.   I. Title.
TH435.T88 1986   692'.5   85-15759
ISBN 0-442-28309-1

For my mother and father,
Kathleen and Jack Turner

Acknowledgments

To the following, for the valuable information they supplied:

Thomas F. Bellows
W.S. Bellows Construction Co.

Bruce M. Grosse
Pyramid Homes

Vincent Kim
Vincent Kim Custom Homes

Earl Thomas
United States Gypsum

Contents

Preface

Before the Industrial Revolution, major buildings in the Western world were essentially composed of one technological component serving a variety of structural, environmental, and functional purposes. Since the onset of industrialization, discrete specialized components have been developed to satisfy the requirements of building. At one time the large building consisted of a masonry structural frame, abetted by foundations and minor amounts of glass and applied finishes. Now a structural podium and frame are made to support an envelope that seals off the interior environment; this environment is modified by machinery that weaves through the web of the frame. All of this is in turn covered by nonstructural infill materials that also serve to create the smaller spaces needed for living and working.

Although the technological changes wrought by industrialization to small-scale residential building types have been less monumental than those to major buildings, they have nevertheless been profound. The preindustrial small building consisted of a masonry or wood frame integrated with its enclosure, sitting on top of a masonry podium. Although applied finishes, particularly plaster and wood moldings, might adorn the interior, the shape of the frame largely defined the shapes and sizes of interior spaces. Today, the wood or masonry frame still sits atop the podium, and the infill still consists largely of applied finishes. Industrialization has changed all else. Despite outward appearances, the envelope is now a discrete veneer, and the interjection of machinery into the interior environment is no less significant than in major buildings. The principal impact of the industrial age upon the podium, frame, and infill has been in material substitutions and erection procedures.

As the technology of building has changed, so has the structure of the

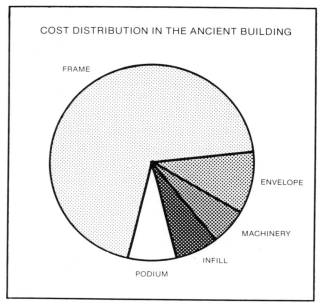

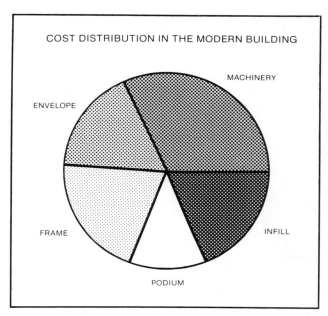

P-1. Cost Distribution in the Ancient Building.   **P-2. Cost Distribution in the Modern Building.**

professions and trades involved in the construction process. In many cases these alterations have changed the technologies as well. We now have specialized design professions for each technological component, as well as specialized construction trades for each. Before the 1800s, master masons or master builders directed crews composed chiefly of masons and carpenters. These master builders were generally responsible for design as well.

Such monumental changes in the way buildings are made have a concomitant effect on the way in which the financial resources for buildings are applied. In fact, decisions regarding economics are more the cause than the effect of changes in technology, and the allocation of moneys to particular parts of the building is the primary basis of architectural style.

An examination of the history of cost allocation in buildings will show the reasons for the development of the major stylistic eras in Western architecture. A further examination of recent cost history will illuminate the probable path of stylistic development in the future.

Historical accounts of the evolution of each building component can be an invaluable asset in estimating the relative cost of each component through the years. Such accounts were compiled by reviewing literature regarding architecture and technology. In addition, engineers, contractors, and materials suppliers were consulted with. Some observations from the author's own experience were recorded. Highlights from these sources are presented in the Appendix. They formed the basis of, and supplied the details for, the accounts of the historical development of building components that follow.

The underlying socioeconomic roots of building technologies and costs will not be at issue here and will only be mentioned to stress particular points. This investigation was approached as a microeconomic study, concentrating on the allocation of monetary resources within the building itself.

CHAPTER 1

The Component Technologies of Buildings

Through the centuries five basic technological components have been used in the construction of buildings. These are the podium, frame, envelope, machinery, and infill. In various historical periods, each has been used to varying extents, and some components once served functions now fulfilled by others.

The *podium* is defined as the base of the building and the devices used for vertical transit. It includes foundations, subgrade floors, the grade-level floor, special at-grade structures that are integral to the below-grade construction, and elevators. This component is foundation- and transportation-oriented.

The *frame* consists of the load-bearing members of the building aside from the podium; these components carry both live (people and furniture) and dead (building weight) loads. It also consists of finishes that are integral to the structural material or are necessary for its structural soundness. Thus, sculpted granite blocks, if load-bearing, are frame members, although the sculpting may not in itself contribute to the strength of the stone. Spray-on fireproofing used on structural steel is also a frame-related cost item. Likewise, plywood sheathing on a balloon-frame house is considered part of the frame.

The *envelope* is the building enclosure that separates the interior environment from the exterior environment. It carries no load save its own. A modern curtain-wall system on a high-rise building and the stained glass that adorned Gothic cathedrals are both envelope items. So is the brick veneer on a residence.

The *machinery* consists of the environmental, sanitary, and life-safety equipment that is routed through the interstices of the frame. Heating, ventilation, and air-conditioning equipment are part of the machinery, as is electrical work. Plumbing, fire-extinguishing, and alarm systems are also included.

The *infill* consists of nonstructural, nonmechanical interior materials used for space-dividing and/or decorative purposes; specialty items are also considered to be infill. Partitions, applied finishes—such as plaster and paint—casework, and fixed equipment help make up the infill. Fixtures, furnishings, and movable equipment are not considered as infill with regard to building cost.

This five-part categorization is based on several factors. First, each component serves a unique purpose in the physical constitution of the building. Second, the materials and techniques used in the construction of each component are different. Third, the complexity of construction on a project today requires a division of labor and a sequencing of construction according to easily identifiable tasks; the purpose and materials of the components help to identify these tasks. Fourth, as the number of people and materials involved in building has multiplied, construction trades and design professions have organized into specialties that make possible proficiency in one particular component.

Building Systems Identified by Cost Estimating Guides

Building Component	Means Systems Costs 1983	Dodge Construction Systems Costs 1984
Podium	Foundations Substructures Conveying	Foundations Floors on Grade Conveying Systems
Frame	Superstructure	Superstructure
Envelope	Exterior Closure Roofing	Exterior Walls Roofing
Machinery	Mechanical Electrical	HVAC Plumbing Electrical
Infill	Interior Construction	Partitions Wall Finishes Floor Finishes Ceiling Finishes Specialties Fixed Equipment

Identification of the podium, frame, envelope, machinery, and infill is not inconsistent with the categorizations for building systems delineated by the principal construction cost-estimating resources in the nation. Systems identified by the R.S. Means Company in its 1983 edition of *Means Systems Costs*, those pinpointed in McGraw-Hill's *Dodge Construction Systems Costs 1984*, and the five components outlined here are compared in Figure 1-1. Note that the cost guidebooks merely take finer slices through the building.

Major buildings are identified here as those whose size makes the use of a wood-frame, load-bearing wall structure impossible. Although this encompasses a wide variety of building types and sizes, they are all analogous with regard to their basic materials and construction techniques. Historically, major buildings have generally been dedicated to civic and religious purposes in classical times; ecclesiastical, monastic, and civic uses in the Middle Ages; and office, industrial, mercantile, and institutional purposes in recent eras.

Small-scale structures are identified here as residential buildings. It is true now, as it has been throughout history, that residences most often use the load-bearing wall structural system, especially if it is of timber.

Evolution of the Component Technologies in Major Buildings

The Podium

Although the basic technology of the podium remained essentially unchanged until the latter part of the nineteenth century, different civilizations made varying use of the spaces made possible by podium construction. The Parthenon in ancient Athens used a simple stone base to distribute the compressive forces transmitted down through the columns into the ground, as shown in Figure 1-2. The foundations of Winchester Cathedral, seen in Figure 1-3, are representative of the Gothic era in the Middle Ages, and show little advancement beyond the technology of the Parthenon. At Winchester, maximum compressive forces are accommodated at points or lines under piers or walls by rubble fill foundations made by filling trenches with rough-cut stones bound in mortar. As in the Parthenon, little use of subterranean space is evident.

Between the Greek and Gothic periods, the Romans exploited their technical aptitude with arches by constructing below-grade spaces, as at the Colosseum. With the fall of the Empire, this technology fell into disuse. It was rediscovered during the Renaissance, after which extensive use of

1-2. The Classical Podium.

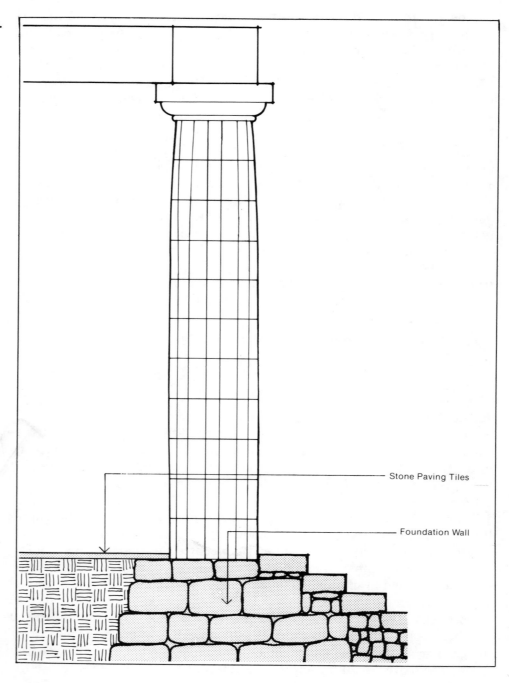

Stone Paving Tiles

Foundation Wall

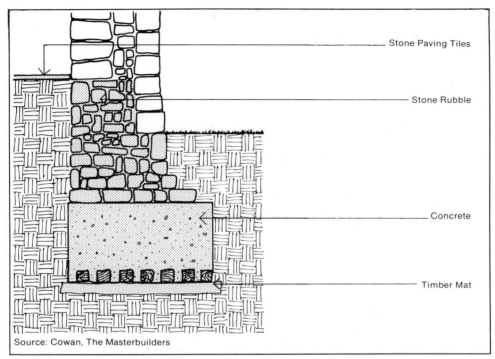

1-3. The Medieval Podium.

Stone Paving Tiles

Stone Rubble

Concrete

Timber Mat

Source: Cowan, The Masterbuilders

1-4. The Renaissance Podium.

Stone Masonry Arch

Stone Masonry Footing

Source: Condit

1-5. The Modern Podium.

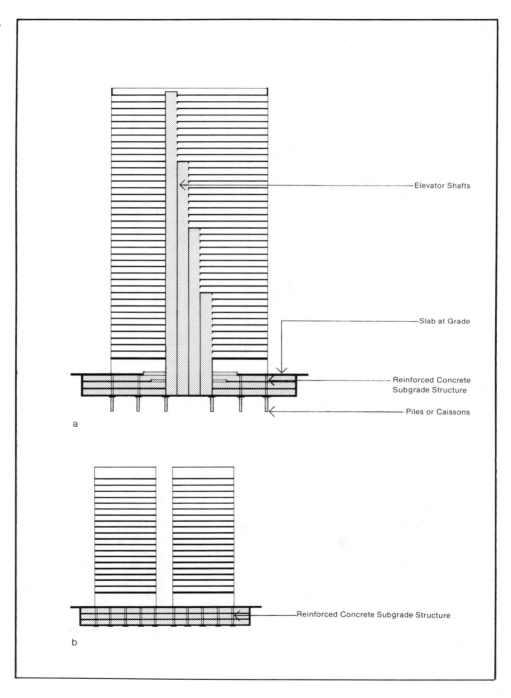

subgrade spaces became commonplace. This can be seen in Figure 1-4, which shows Founder's Hall at Girard College in Philadelphia.

In Figure 1-5 two examples of modern podium technology are shown. The effects of reinforced concrete are clearly visible. The podium is now a rigid structural skeleton that transmits the loads from the building above into the soil below. The mass of material used in earlier ages is gone; subgrade floors are virtually as open as the typical office floor. Only the grade-level slab is massive, since it is used as the staging area for construction and thus must hold trucks, stored materials, and so forth.

As significant as the introduction of reinforced concrete was, the invention of the elevator was even more consequential. The development of high-rise buildings, made possible by elevators, has affected the podium. Special structural accommodations are needed at grade level, and these are often taken care of in the construction of the podium. Such accommodations are needed for the major interchanges of the machinery of the building—horizontal runs, usually at grade level, connect the major mechanical, electrical, and plumbing equipment below with the vertical chases that supply services to the building above. Such accommodations are also needed because the ground level is normally the interchange between the vertical transportation within the building and the ground-level transportation outside. Thus, the podium may be the staging point for the movement of large numbers of people. Since the subterranean levels are often used for parking or other secondary uses, the grade level may also serve as the interchange between two different occupancy types within a building. Figure 1-5 illustrates the podium as the staging point for the vertical transit of people in the building; elevator routes are clearly visible.

In summary, the podium of a major building has undergone a transformation from a simple underlayment to distribute the structure's weight

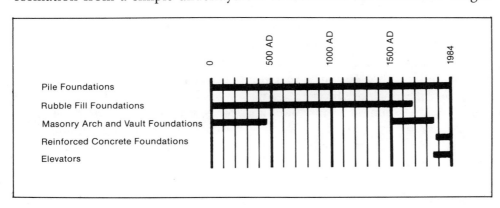

1-6. Evolution of the Podium.

into the earth to a multilayered open frame that now distributes not only loads but people as well. Major technological systems used in the podium are illustrated in the historical graph in Figure 1-6.

The Frame

One of the components most radically altered by industrialization was the structural frame. Until the mid-nineteenth century, load-bearing masonry construction was the only technology available for large buildings. On occasion, other materials worked in tandem with masonry. For example, iron chains have been used at least since the Renaissance, iron clamps were known to the Romans, and other metal devices were used to reinforce masonry construction even earlier. Construction before the steel age, however, depended primarily on the compressive strength of stone masonry to make a building frame.

In addition to acting as the structural support of the building, the frame was also the enclosure for the interior environment, excepting the small areas taken up by glass and doors. The frame provided divisions between different interior spaces as well. The sheer mass required in masonry buildings also helped to modify the interior environment, since the extremes of temperature on the outside were not transmitted rapidly through the thick walls.

A large expense in masonry construction was carpentry. This was due to the large amount of wood centering, or formwork, that was erected to enable workers to place the stones until an entire arch was completed. This is similar to the construction of modern reinforced concrete structures, where carpenters still build formwork.

Although the frame was essentially the same until the 1850s, some variety did exist from age to age. The Romans built stone arches, which they used as a permanent formwork for concrete construction. With a ready supply of slave labor and codes that prescribed death as the penalty for building collapses, Roman officials ensured that structures would be massive. After the fall of the Empire, churches were virtually the only major buildings constructed in the West. Since they doubled as places of refuge during these unstable times, windows were frequently omitted and walls were thick. Construction techniques and building forms still imitated the Roman in this, the Romanesque period of architecture. Evolution of the frame in the latter part of the Romanesque era is shown in Figure 1-7. Massive masonry walls resisted the thrust of the barrel vault. Eventually, small openings were introduced into the walls.

Gothic architecture dominated Western civilization in the Middle Ages.

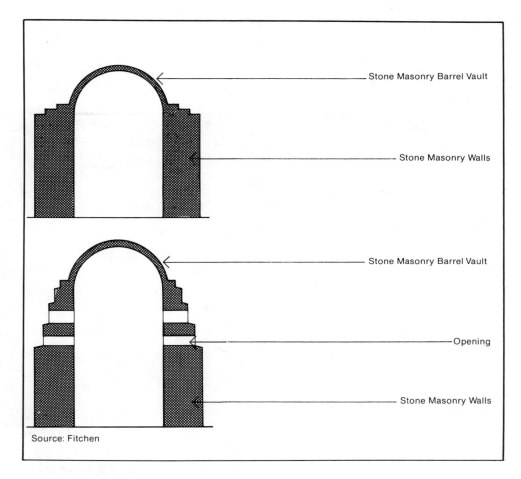

Stone Masonry Barrel Vault

Stone Masonry Walls

Stone Masonry Barrel Vault

Opening

Stone Masonry Walls

Source: Fitchen

1-7. The Romanesque Frame.

Builders of this period made a few improvements on Romanesque building techniques that enabled the frame to shed some of its mass. Eventually, the frame became a masonry skeleton, its interstices filled with stained glass. Recurrent labor and material shortages and the growth of a continental glass industry provided the impetus for these developments. Compare the bulk of the frame at Chartres Cathedral (Figure 1-8), the epitome of Gothic building, with the Romanesque frame. The pointed arch, groin vaulting, and flying buttresses skeletonized the frame and enabled the envelope to expand.

Interest in the classical world was reawakened in Europe, and eventually in the United States, by the Renaissance. Domed masonry construction was adapted to the buildings of the day, shed of the bulk of Roman and Roman-

1-8. The Gothic Frame.

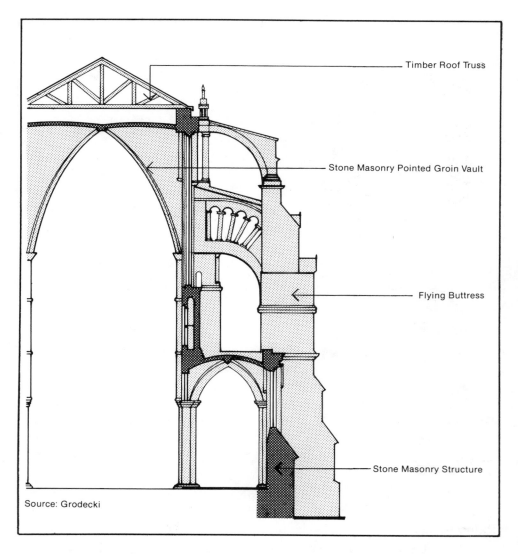

Timber Roof Truss

Stone Masonry Pointed Groin Vault

Flying Buttress

Stone Masonry Structure

Source: Grodecki

esque times. In the Roman Pantheon of 123 A.D., there is one masonry dome whose thrust is counteracted by massive masonry walls. Renaissance builders, however, benefited from the techniques for structural efficiency learned during the Gothic era and added a few innovations of their own. Multiple-shell domes and iron chains that opposed the outward thrust of the domes are examples. As a result, although ancient building forms reappeared, building mass did not appreciably increase from the Gothic. St. Paul's Cathedral in London, designed by Christopher Wren, is perhaps the

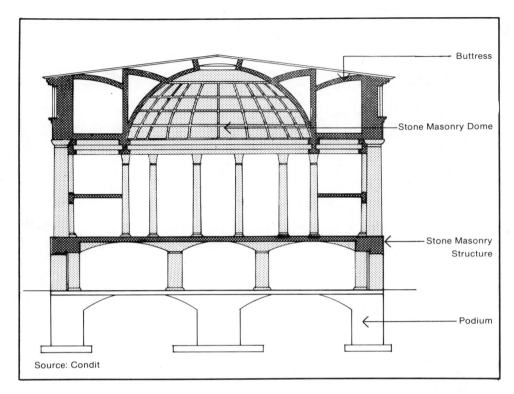

Buttress

Stone Masonry Dome

Stone Masonry
Structure

Podium

Source: Condit

1-9. The Renaissance Frame.

best illustration of this point. Begun in 1675, its dome was triple-shelled, with masonry inner shells and a lead-covered timber outer shell. This is similar to Gothic nave construction, with its masonry vaults and timber roof. St. Paul's Renaissance facade is also supported by concealed flying buttresses. Figure 1-9 shows the United States Sub-Treasury Building, erected in the early 1800s, another example of Renaissance-era construction. Ironically, it is nearer the technologies of ancient Rome than those of the end of the century in which it was built.

One of the most significant occurrences in the history of building took place in London in 1851. In that year Joseph Paxton designed and constructed the Crystal Palace. An exhibition hall made completely of prefabricated iron and glass pieces, it was the first significant structure to divorce the structural frame from the building envelope entirely. No longer did the frame keep out the elements or enclose interior spaces. It was an independent component used solely to support the other components of the building.

Monumental as it was, Paxton's achievement was not fully appreciated for some time. Iron had been used in many buildings before the Crystal

1-10. The Modern Frame.

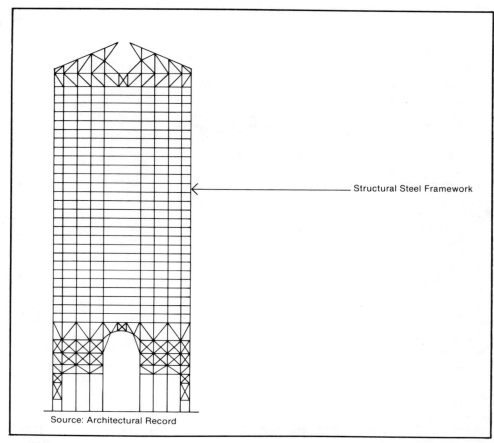

Structural Steel Framework

Source: Architectural Record

1-11. Evolution of the Frame.

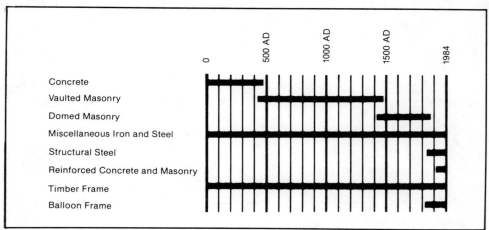

Palace, and iron and steel would be used commonly thereafter. It was not until the twentieth century, however, that the conception of a building as a set of independently manufactured and assembled pieces, each with a specific function, would be accepted. By that time, architectural tastes had changed, problems in protecting iron or steel structures from fire had been solved, and advances in envelope technologies were finally being exploited by designers and builders.

Figure 1-10 illustrates the AT&T Building—a recent example of a structural frame that is an independent component in the building. In this building, the steel frame provides all structural support for the building's component parts. Interesting as the frame's appearance is, it is in no way visible and is hardly reflected in the architectural expression of the building. Other components cover it.

In summary, the frame has evolved from an exposed massive stone structure that served many purposes to a steel or reinforced concrete skeleton with none save the structural function. The non-load-bearing purposes that the frame once fulfilled have been assumed by an expanding envelope, machinery, and infill. Major technological systems used in the construction of the frame are illustrated in the historical time chart in Figure 1-11.

The Envelope

Historically, the importance of the envelope as an independent component in the building has depended on several factors, not the least of which is the technology of glass. Although glass was known to the Romans, it was little used in their buildings. After the collapse of the Empire, glass all but disappeared until its resurgence in Venice in the Middle Ages. Stained glass flourished during the Gothic period of architecture, to be superseded by the development of crown glass during the Renaissance. The visual advantages of clear glass in large (by previous standards) panes doomed the use of stained glass, and crown glass made this possible. In the late nineteenth century, industrialization revolutionized the manufacture of glass, and large sheets of clear plate glass, such as those used at Paxton's Crystal Palace, became commonplace.

As influential as glass technology was, the evolution of the frame probably had a more immediate impact on what happened to the envelope in the modern age. Before iron and steel were mass-manufactured in structural shapes, there was no real impetus for the development of independent envelope technologies such as plate glass. As long as the frame remained massive, it assumed the role of an enclosure.

Other important factors in the development of the envelope were sheet metal, insulation, and exterior stone veneer technologies. These evolved as builders searched for lighter and more thermally resistant enclosures for the structure.

Prior to the industrial age, the principal means of architectural embellishment was the shaping of the mass and surfaces of the masonry structure. Glass was used only as an interstitial filler between structural members and became a decorative feature itself. One may trace the relative importance of glass, and hence the entire envelope, in Figures 1-7 to 1-9. The degree of prominence of the envelope has historically been inverse to the mass of the frame.

When steel and iron began to see extensive use in the structure of the building, architectural style did not immediately adapt. In the era of the historical revival styles of the late 1800s, masonry envelopes were constructed to resemble the structural frame common in preindustrial times. Ironically, glass was used as an interstitial filler between other envelope elements that simply gave the illusion of being frame members.

In the twentieth century, architects can design building enclosures that exhibit little relationship to the frame at all. The envelope is now an independent structure attached to the frame with point connections of miscellaneous steel. These connections may be in various locations on the columns or perimeter beams of the building. Continuous fire separation between floors is needed to seal the gap between the floor edge and the back side of the envelope material.

One result of this technology is that the envelope has become akin to a mask—similar to those used in earlier cultures—that projects an image of ritual, social, or aesthetic significance, while satisfying minimal visual or ventilation needs for those behind it. Stained glass and statuary accomplished these ritual and other functions in previous ages. Today, the AT&T building shown in Figure 1-12 is an example of the possibilities inherent in modern "mask" technology.

In summation, the envelope has evolved from glass used as a filler between the massive frame elements to a thin veneer that may completely wrap the insides of the building, giving little or no indication of the physical makeup of the building within. Historical developments in the technology of this building component are illustrated in Figure 1-13.

The Machinery

The most significant development in the history of buildings has undoubtedly been the introduction of electrical and mechanical devices that actively control and alter the interior environment. Additionally, advances in plumbing equipment and fixtures have made possible the extension of sanitary systems to the farthest reaches of buildings. Moreover, life-safety measures have recently (in the past fifteen years) been incorporated into buildings; sprinkler and alarm systems are now regular features in large structures. Today, the machinery is likely to be the single most expensive component in the building.

Prior to the nineteenth century, plumbing was the only real environmental machinery, and its cost was relatively insignificant. Water supply and drainage existed in Rome but virtually disappeared as the Empire disintegrated. Water and sanitary systems were present in some medieval buildings, such as monasteries, but they were limited and primitive. It was not until major public works projects were undertaken by the British, French, and other continental monarchies that an extensive public water supply began to reappear in the 1700s. In the latter part of the eighteenth century the water closet was invented, heralding the birth of the modern toilet room with its various sanitary accommodations. Modern plumbing systems were basically "invented" by the middle of the 1800s.

Before 1900 thermal environments in buildings were actively modified by a few means. The hearth or fireplace was used for heating interior spaces, as were cast iron boilers—an innovation of the 1700s. Both were also used for food preparation. Since there is no reliable way to distinguish the percentage of the cost of a masonry fireplace that is attributable to nonstructural purposes, its cost is included wholly in the frame's. And since cast iron boilers were primarily used for cooking, they are more correctly considered appliances, not building-cost elements.

Until this century, lighting in buildings had always been provided by fire. As late as 1817, major buildings in the United States were still using candlelight. In the early 1800s, gas lighting became widespread in England. Gas became an economical and popular means of lighting in Europe and in the U.S. and was a strong competitor to electricity until the second decade of the 1900s.

Building science was revolutionized by the invention of electricity, more precisely by the domestication of electric supply. This occurred first in 1882 through the efforts of Edison in New York and concurrently by Swan in London. Reyner Banham, in *The Architecture of the Well-Tempered Environment*, has pointed out the almost immediate escalation of electricity use from that year on.

Building technology was further transformed by modern Heating, Ventilating, and Air-Conditioning systems (HVAC) that were pioneered by Willis Carrier in this country. Although the fundamentals of HVAC were well developed by Carrier as early as 1906, and fully air-conditioned buildings

1-12a. The Modern Envelope.

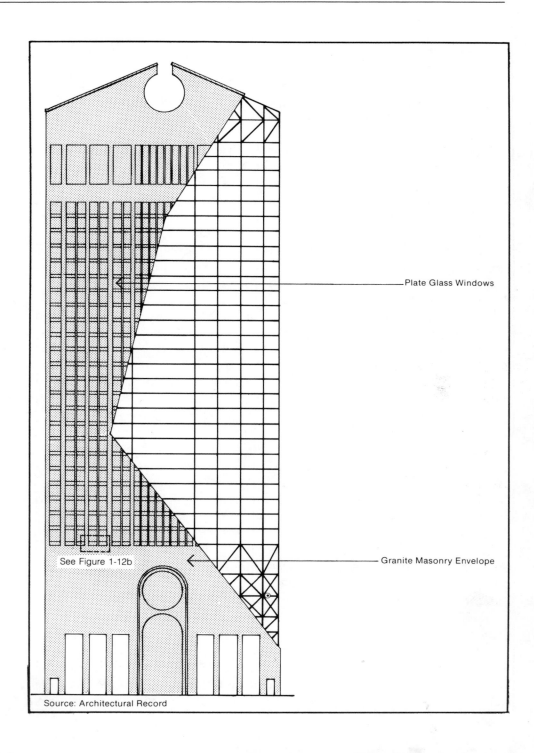

Plate Glass Windows

Granite Masonry Envelope

See Figure 1-12b

Source: Architectural Record

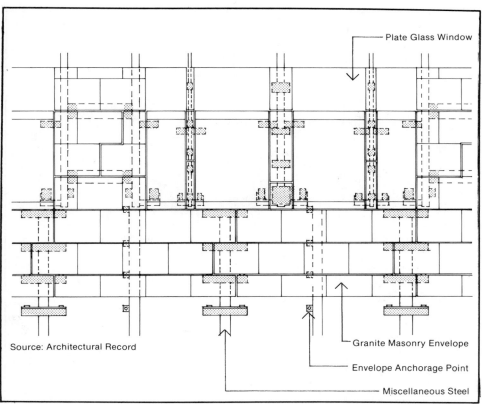

1-12b. The Modern Envelope.

Plate Glass Window

Source: Architectural Record

Granite Masonry Envelope

Envelope Anchorage Point

Miscellaneous Steel

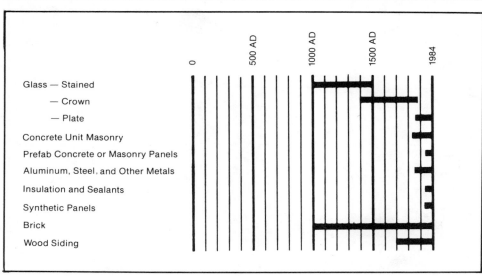

1-13. Evolution of the Envelope.

	0	500 AD	1000 AD	1500 AD	1984
Glass — Stained					
— Crown					
— Plate					
Concrete Unit Masonry					
Prefab Concrete or Masonry Panels					
Aluminum, Steel, and Other Metals					
Insulation and Sealants					
Synthetic Panels					
Brick					
Wood Siding					

1-14. Environmental Controls Prior to Industrialization.

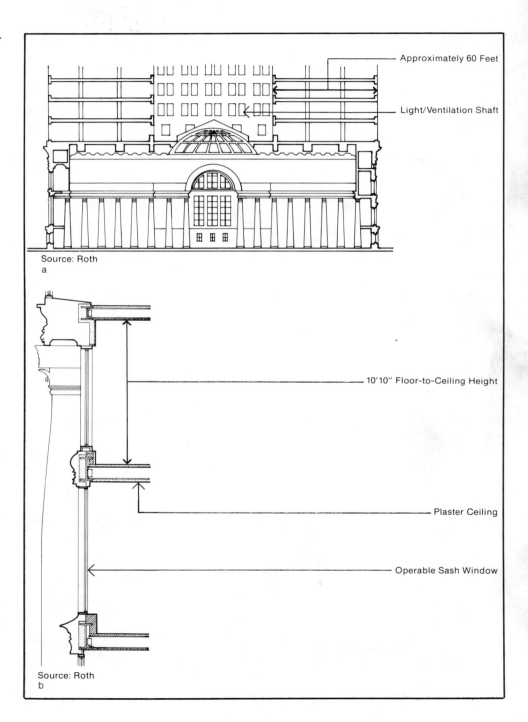

Approximately 60 Feet

Light/Ventilation Shaft

Source: Roth
a

10'10" Floor-to-Ceiling Height

Plaster Ceiling

Operable Sash Window

Source: Roth
b

were constructed in the 1920s, it was not until after World War II that complete climate control inside buildings was to become commonplace.

Figure 1-14 shows buildings in the stage between the introduction of electricity and the development of HVAC. Gas light and then electric light made larger floor areas in buildings possible, since natural light was no longer so crucial for illumination. Ceiling heights remained high, in order to facilitate air movement. Ventilation shafts penetrated the cores of large buildings to allow cross-ventilation inside. Light fixtures were attached directly to the surfaces of walls and ceilings in much the same manner that candelabra and chandeliers had been mounted in the past. The National City Bank building in Figure 1-14(a) was completed in 1909 but hardly represents twentieth-century technology. Plumbing for water supply and sanitary disposal was the only element of modern environmental systems present. Lighting was probably achieved with pendant-mounted incandescent fixtures. The ceiling was attached directly to the underside of the floor structure, creating interior office spaces nearly eleven feet high, thus expediting air movement. The New York Municipal Building, shown in Figure 1-14(b) and built in 1908, is also an example of this.

The impact of modern machinery systems may be seen in Figure 1-15. Systems are distributed to individual locations on each floor from vertical shafts that are connected to the major equipment. HVAC systems are driven by equipment in a central plant, usually at or below grade level, or in a rooftop penthouse. From these major facilities the climate control system is distributed through the central shaft into the plenum formed by the underside of the structural floor and the ceiling suspended from it. Air-handling units on each floor may facilitate this. In order to increase the efficiency of the HVAC system, it is currently desirable to minimize the ceiling height, thus reducing the volume of space that needs conditioning. Electrical wiring is routed from equipment in a vault near ground level up through separate shafts to electric panel rooms on typical floors. From here it is distributed out to the power outlets and lighting fixtures. Communications equipment follows much the same path. Plumbing and fire-extinguishing systems go up the building in risers, from which they branch out to the individual fixtures on each floor. Figure 1-15a clearly shows the mechanical penthouse. In Figure 1-15b, the suspended ceiling in the John Hancock Tower in Boston forms a space below that is only about 8′6″ high.

In summary, buildings have been changed radically by the advent of systems that control the interior environment. In less than a century we have seen passive means, which at best modify the extremes of exterior conditions, give way to the sealing and active alteration of interior spaces.

1-15. The Machinery in the Modern Building.

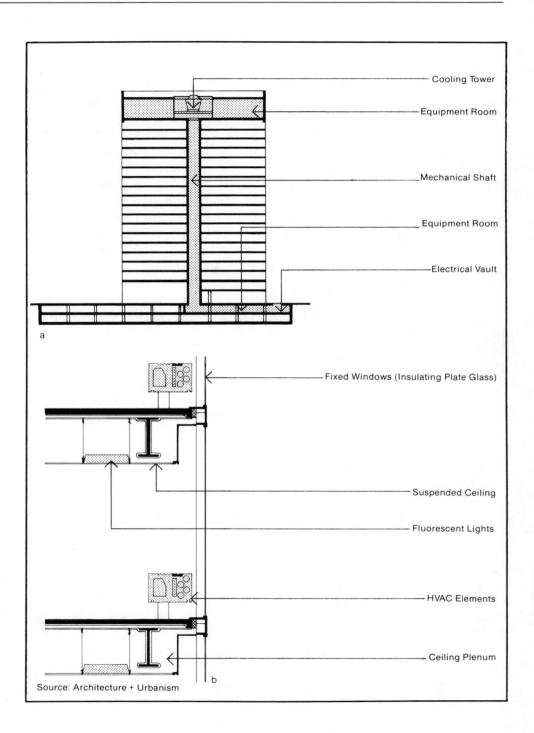

Cooling Tower

Equipment Room

Mechanical Shaft

Equipment Room

Electrical Vault

a

Fixed Windows (Insulating Plate Glass)

Suspended Ceiling

Fluorescent Lights

HVAC Elements

Ceiling Plenum

b

Source: Architecture + Urbanism

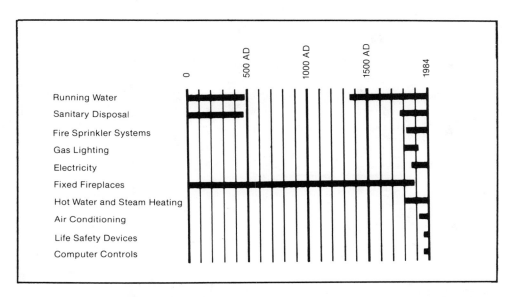

1-16. Evolution of the Machinery.

More recently, building code requirements have made fire-extinguishing (sprinkler) systems and fire alarm systems common. The historical development of machinery systems in buildings is illustrated in Figure 1-16.

The Infill

Not until the first "revolution" of the nineteenth century—the introduction of steel—had occurred, did the utilization of nonstructural materials for decorative and space-dividing purposes become widespread. Unlike the transformation of the frame, the development of the infill was a following, not a leading innovation. And unlike the machinery, the infill represented less a watershed in the advancement of technology than the application of modern techniques to age-old construction methods.

The use of filler materials to divide space and create visually interesting finishes has been a well-known technique for centuries. Vitruvius gave detailed instructions in the use of wattle-and-daub, the precursor of the modern drywall system. He also wrote about plaster and paint. In major buildings, however, the extent of these materials was likely to be relatively minimal, except for painting. Since the masonry frame was so massive and its elements so closely spaced, little need existed for a space-dividing technology independent of the structure. This may be seen in the Roman Palazzo Farnese (designed by Antonio de San Gallo the Younger in 1534), illustrated in Figure 1-17. Although plaster was applied to interior surfaces even during the Middle Ages, it was not fully resurrected until the late Renaissance,

1-17. The Infill Prior to Industrialization.

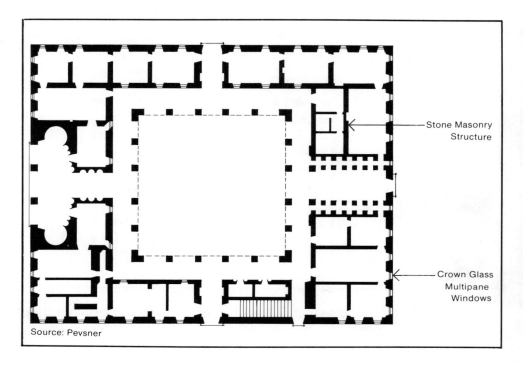

Stone Masonry Structure

Crown Glass Multipane Windows

Source: Pevsner

during the baroque and rococo periods. In these periods, flamboyant expressions in plasterwork were common, especially in the ecclesiastical architecture of Mediterranean nations in Europe. Such plasterwork, as at the Charterhouse in Granada, Spain (ca. 1700), may be seen in Figure 1-18. It was applied to a backup that concealed the true configuration of the structure. Plaster was often painted to resemble stone masonry, with exquisite veining and jointing. In England, plaster came to enjoy popularity as well. King Henry VII granted a charter to the "plaisterers" guild in 1501, an indication that this material had moved beyond use simply as a wall coating.

Prior to plaster's revival in the Renaissance, interior ornamentation had been achieved by the surface-shaping of the stone frame. In this respect, decoration had been more akin to the formed concrete or decorative concrete masonry units of the twentieth century.

When the introduction of steel to buildings accelerated the development of the skeleton frame, the need arose for some element that would partition spaces. Technologies already existed that could fill this requirement. Before 1800, an adaptation of wattle-and-daub had developed in North America. Wattle-and-daub consisted of vertical wooden stakes with straw between them, all covered with plaster or mud. By substituting brick for the straw

1-18. The Baroque Infill.

Plaster

and then covering the stakes and bricks with clapboards to reduce thermal stress in the assembly, American colonists "invented" the wood stud wall system. The brick was eventually eliminated, and advances in milling lumber coupled with the mass production of iron nails in the early 1800s made possible a lightweight partition system of standard-size, easily transportable elements.

Despite its early wide use in residential and other small construction, the stud wall system was not used effectively in major buildings until well into the twentieth century. The combustible nature of wood and the expense and complicated construction process involving plaster mitigated against this partitioning system. With the invention of Sackett-board, later to become gypsum wallboard, the problems with plaster were overcome. In the 1960s, the appearance of the screwable metal stud, the screw gun, and the self-drilling, self-tapping screw overcame the last obstacles to stud wall construction in major buildings.

After the mid-1800s and until the 1960s, partitioning was largely accomplished through the use of prefabricated masonry units. Concrete blocks first appeared in the 1830s, and other masonry units, such as gypsum wall tiles, also came on the scene. The relative ease of transporting and assembling these standard units made them popular. In high-rise structures, masonry units were fairly easy to lift to the upper stories during construction. These units could be made of lightweight aggregate and with hollow cores in order to reduce their weight.

Ceilings continued to be made of plaster after the onset of industrialization. The lath would be attached directly to the underside of the floor structure. Suspended ceilings that concealed mechanical equipment were used as early as 1906, but did not become widespread until the acoustical tile ceiling appeared in the 1930s.

After World War II, "open-plan" office space became popular. The reasons for this were twofold. First, masonry partitioning, both in itself and in the structure required to support its weight, was expensive. Second, the acoustical ceiling and broadloom wall-to-wall carpeting solved the noise problems that the open office plan posed.

Figure 1-19 shows the possibilities inherent in the skeleton frame, where the structure plays no part in dividing spaces. This example is once again from the AT&T Building. In fact, designers of today's office buildings seek to eliminate columns from all floor areas but the core and perimeter; the entire office floor may be devoid of space-dividing elements. Figure 1-20 illustrates the interior of the General Motors Technical Center, an example of an open-plan space with acoustical ceiling.

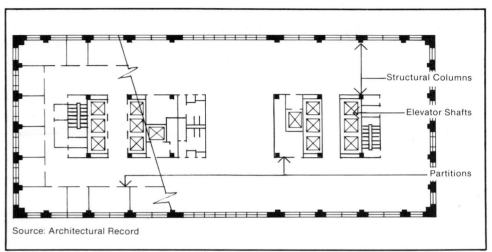

Source: Architectural Record

1-19. The Infill After Industrialization.

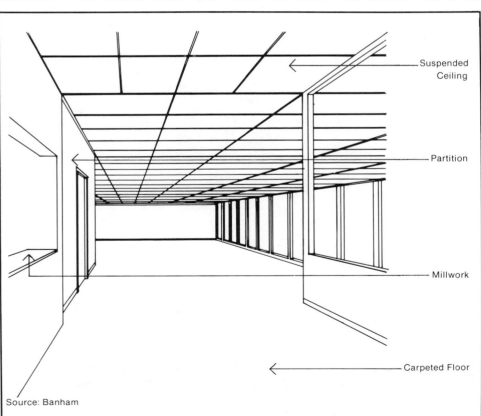

Source: Banham

1-20. The Modern Infill.

1-21. The Modern Infill.

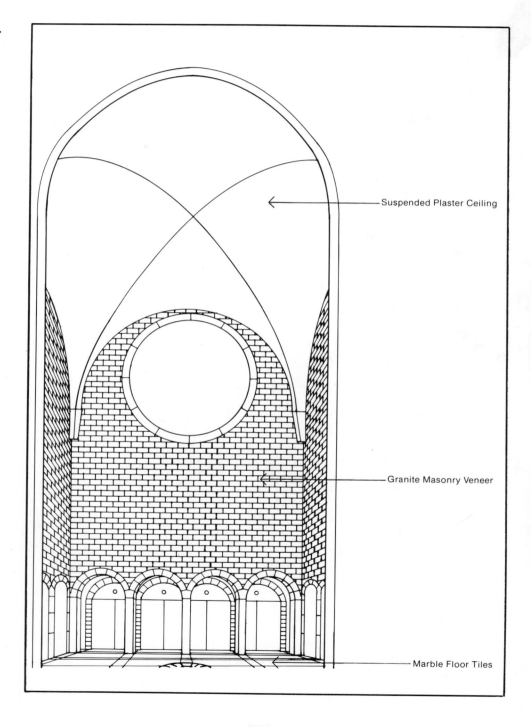

Suspended Plaster Ceiling

Granite Masonry Veneer

Marble Floor Tiles

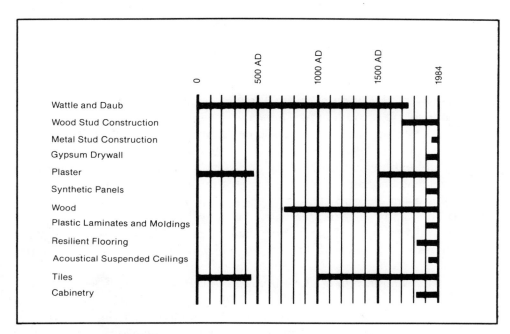

1-22. Evolution of the Infill.

In the twentieth century, infill technology has been explored and refined. Thin veneers for ceilings, floors, and walls have been developed, made possible by advancements in modern adhesives. Plywood, particle board, plastic laminates, metallic laminates, resilient flooring, and other such now-common materials were practically unheard of until the 1900s. Stone masonry veneers are also common, made possible by advancements in stone-cutting machinery, steel anchorage devices, and high-strength mortars. The infill serves as a cover for the podium, frame, and machinery, which are erected quickly and crudely, without much attention to appearance. In Figure 1-20, virtually everything visible to the naked eye is infill, including suspended ceiling, floor covering, and partitions. Despite the massive appearance projected in the lobby of the AT&T Building, shown in Figure 1-21, this entire space is actually a pastiche of thin veneers anchored to the building's frame.

In summation, infill technology has existed for as long as mankind has desired to conceal the rough edges of construction with decorative features. About a century ago, the infill had to assume the function of partitioning interior spaces as well. Since then, the infill has become not unlike the envelope—a thin facade behind which the building hides. The building industry has been striving to make this facade as thin and lightweight as is feasible. Figure 1-22 gives a historical account of the evolution of this component.

Evolution of the Component Technologies in Small-scale Buildings

The Podium

The history of the podium in small-scale residential buildings differs little from that in major buildings. The small structure has seldom made use of piles and other sophisticated underpinnings, although the Italian Renaissance architect Andrea Palladio specified them for his monumental homes and villas. Rubble fill was the mainstay of residential foundation work until the late nineteenth century. Foundations of the Etruscan house unearthed at Marzabotto (ca. second century B.C.), shown in Figure 1-23, consisted of such rubble, and the floor area between the walls was of compressed earth; exterior walls sat upon the stone foundation walls. This is essentially identical to Vitruvius's description of the podium, which followed a couple of centuries later.

As with major buildings, different civilizations made varying use of subgrade construction. In few instances, however, is there evidence of below-ground spaces. One example of such is Thomas Jefferson's Monticello (1770–78). Many expensive homes on the European continent also had

1-23. The Classical Podium (Residential Buildings).

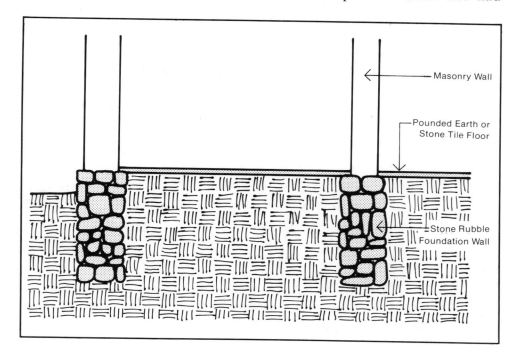

Masonry Wall

Pounded Earth or Stone Tile Floor

Stone Rubble Foundation Wall

subgrade spaces, which were used for wine cellars, among other purposes. Palladio, who so influenced Jefferson, made no mention of subterranean usage, however.

It was not until the late 1800s that residential foundations were significantly altered. Wood, concrete, and steel technologies combined to make this possible. As will be noted when the frame is discussed, the wood balloon-frame house came into its own in the mid-nineteenth century. This lightened the overall weight of the structure considerably and obviated the need for the massive masonry foundations that characterized houses of earlier eras. For buildings intending to use subterranean space, the major loads acting on the podium thus shifted from the compressive forces from the frame above to the loads from the earth acting laterally on the perimeter walls. The ideal solution to podium design was now to create a retaining wall around the perimeter of the house and to pick up intermediate loads with small columns situated as needed beneath the wood framing of the first floor. The columns would in turn transmit loads down to spot footings beneath the cellar floor. Concrete and steel technologies made this possible. Monolithically poured reinforced concrete or prefabricated concrete masonry units made possible a thin perimeter wall, and prefabricated steel columns, round or rectangular in section, could handle the intermediate loading. As a result, the cellar became open and spacious; this may be seen in Figure 1-24. In a home of this sort, the first-floor wood structure is considered part of the frame.

Although the podium described above shows the most evidence of the changes introduced in the industrial age, the cellar in our age is generally used only in areas with cold climates, where outdoor activities are not feasible for a good part of the year. The second generic type of the present-day podium is shown in Figure 1-25. It is used where subgrade space is not desired. This also makes good use of reinforced concrete technology and is called the slab-on-grade with grade beams. It consists of a thin reinforced concrete slab cast monolithically with downturned beams at the perimeter and at other crucial load-bearing points. The wood frame, as well as floor finishes, are attached directly to the slab. The grade beams will typically go down to the frost line. This prevents the disruption of the foundation from the heaving of the soil as it freezes and thaws.

In summary, significant changes in the residential podium did not occur until lightweight wood framing, concrete, and steel made them possible in the 1800s. At that time, rubble walls and floors at grade, or cavernous masonry cellars, began to give way to the concrete retaining wall or slab-on-grade foundation.

1-24. The Modern Podium—Subgrade Space Desired (Residential Buildings).

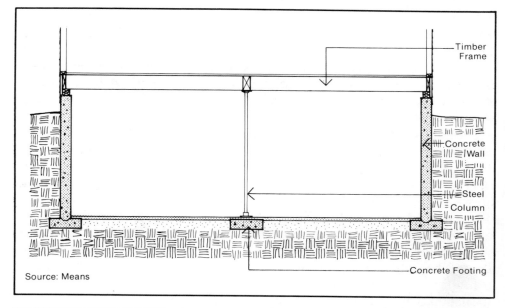

Source: Means

1-25. The Modern Podium—Subgrade Space Not Desired (Residential Buildings).

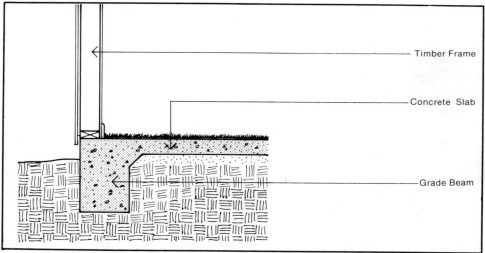

The Frame

As with the frame in major buildings, the residential superstructure underwent profound modifications during the process of industrialization. Unlike what happened with major edifices, this occurred despite the fact that the primary frame material—wood—was not replaced.

Perhaps it is best to consider the sudden alteration to the wood frame after the 1840s, embodied in the balloon frame, as the culmination of a centuries-long process of refinement of wood construction technology. This evolution from wattle-and-daub to the wood stud system has been explained above. Let it suffice to say here that the standardization of wood milling practices and the mass manufacture of inexpensive nails made practicable the construction of a lightweight wood diaphragm frame, covered with a thin, weather-protective membrane—thus the "balloon" frame. This came about in approximately 1850, and thereafter most homes, even many monumental ones, have been constructed in this manner.

Prior to the balloon frame, the residential builder had two choices: the heavy timber frame or the masonry frame. The timber-framed house is illustrated in Figure 1-26. It was a system learned by trial and error, with members generally oversized. (Because there were no test laboratories to confirm the stresses that materials could withstand under loads, builders sized members according to rules-of-thumb, based on experience with spans and member sizes.) It was also a system in which construction of the timber-to-timber connections was tedious and exacting. Doweling and mortise-and-tenon were the common connection techniques until bolts and screws made their debut in the Middle Ages. Structural connections were minimized, resulting in a post-and-lintel frame, between whose members an envelope of wattle-and-daub or brick nogging might be placed. The weight of such envelope materials, in addition to the weight of what might have been a lead roof, also made the use of light wood members impossible. This was a wood skeleton that prefigured the iron frames of a much later age, by divorcing to some extent the functions of exterior closure, environmental modification, and space division from that of structural support.

Until the advent of the balloon frame, stone or brick masonry was also used to some degree for the superstructure. This was especially true in monumental residences or villas, but in some civilizations, such as ancient Rome, Renaissance Italy, and Georgian England, the availability of materials, rather than the size of the home, dictated the structural system. In Figure 1-27 Palladio's Renaissance-era home for Monsignor Almerico, located near Rome, is illustrated. It is a stone masonry, load-bearing structure with vaulted ceilings and timber roof framing. The structure is similar to the larger building shown in Figure 1-9. Unlike Figure 1-26, the frame here serves nearly the same purpose as in a major building of the same period—i.e., its mass provides environmental modification, its substance provides exterior enclosure, and its shape forms the interior spaces.

In Figure 1-28 two differences between early timber framing and the modern balloon frame are apparent. First, the post-and-lintel construction

1-26. The Medieval Frame (Residential Buildings).

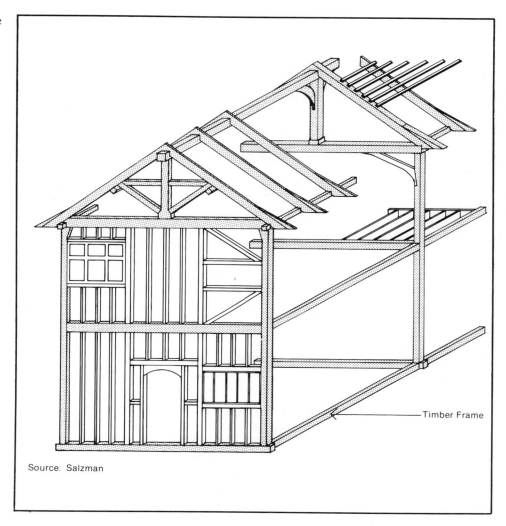

Timber Frame

Source: Salzman

exemplified in Figure 1-26 has been replaced by a diaphragm of many smaller interconnected parts. The milling of lumber in standard shapes and sizes has made this possible by making each structural element in a consistent dimension. Perhaps more important, the introduction of nails has made of the formerly time-consuming chore of wood connections a task that is very quickly and easily performed by even the least skilled of laborers. This has made the wood frame so economical that even the monumental home is now likely to be built with this type of construction. The monumentality, as will be examined, is in the image projected by the masonry veneer

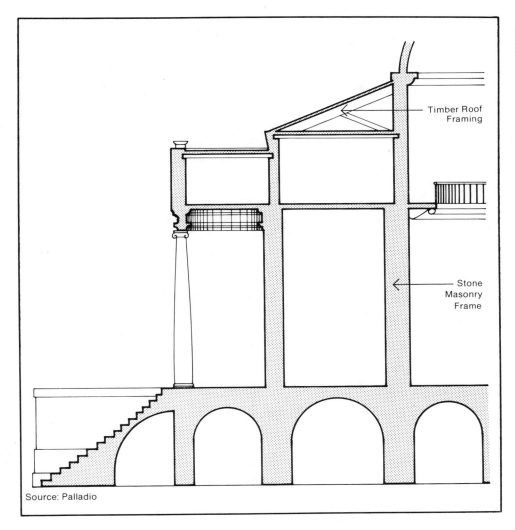

1-27. The Renaissance Frame (Residential Buildings).

Timber Roof Framing

Stone Masonry Frame

Source: Palladio

envelope that is attached to the frame.

In summation, the residential frame, until very recently, has been either of wood or masonry, and the use of masonry has largely disappeared. The wood frame has evolved from a heavy timber skeleton to a light, nearly continuous diaphragm made of interchangeable wood parts. Ironically, this is somewhat the opposite of the evolution of the frame in major buildings, which have gone from masonry "diaphragm" to steel skeleton. A further irony is that steel, in the form of the nail, has generated this evolution in wood construction.

1-28. The Modern Frame (Residential Buildings).

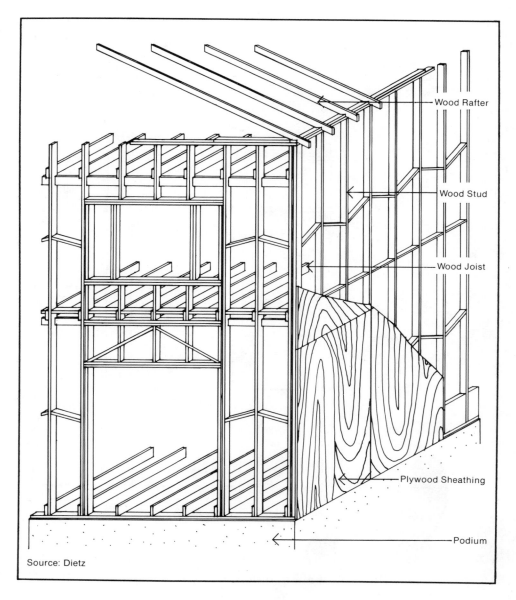

Wood Rafter

Wood Stud

Wood Joist

Plywood Sheathing

Podium

Source: Dietz

The Envelope

As in the case of major buildings, the envelope's importance in small-scale structures has been closely related to the evolution of the structural frame. But its importance has not depended so much on the historical circumstances as on the material chosen for construction of the frame. The mass of the load-bearing masonry, augmented by the small windows, served the purpose of the envelope in preindustrial masonry homes. In timber buildings, the envelope, whether it was wood or masonry, has always existed as a discrete component apart from the frame.

Among the earlier wood frame buildings of some degree of sophistication in Western civilization were those in medieval Europe. Figure 1-29 illustrates an example of the envelope that developed in this period. Brick nogging was placed between the frame members. Often it was covered with plaster, resulting in the look so often imitated in so-called "Tudor" and similar style houses today. Shutters and small windows were often framed out in the walls as well. This is the building system that the English colonists brought to the New World and which evolved into the balloon frame. This envelope was an independent building component, as discrete as any of its modern counterparts.

Homes with masonry frames showed little difference in terms of the extent of the envelope from ancient times to the 1800s. Small openings were typically punched through massive stone walls, as may be seen in Figure 1-30, which shows another Palladio design from the 1500s—the

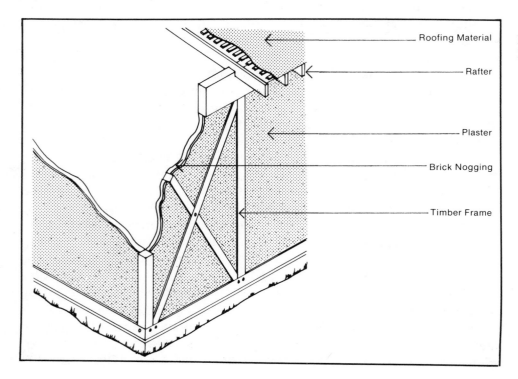

Roofing Material

Rafter

Plaster

Brick Nogging

Timber Frame

1-29. The Medieval Envelope (Residential Buildings).

1-30. The Renaissance Envelope (Residential Buildings).

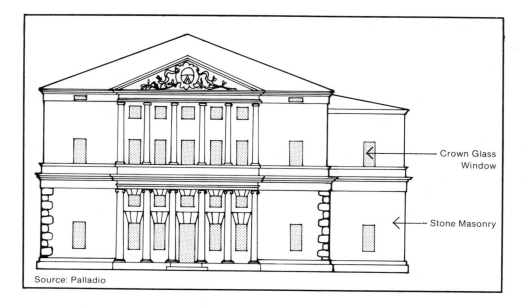

Crown Glass Window

Stone Masonry

Source: Palladio

Antonini home in Udine, Italy. In Rome, such openings would generally be shuttered, but beginning in the Middle Ages improvements in glass technology made windows a more commonplace feature, especially in the homes of the affluent.

In small structures, the roof is the most prominent element of the envelope, and here there has been remarkably little change even to this day. In all homes through the ages, from Vitruvius's to Palladio's to our own, roofs in even the most expensive of homes have been constructed of timber, with a layer of thatch, tiles, shingles, or metal to form the protective membrane. Where wood lath attached to the rafters once formed the substrate for attaching the roofing, plywood sheathing is now likely to be used. Today, layers of water-resistant felt are laid on top of the sheathing before the roofing material is applied. Although thatch roofs have largely disappeared in our civilization, tiles, sheet metal, and shingles are just as likely to be used today as they were in ancient Rome. Industrialization, however, has made it possible to imitate these with less expensive and lighter materials. Thus, concrete tiles are likely to be used instead of clay; asphalt composition shingles are often used instead of wood; and sheet metal—aluminum or terne—has replaced lead.

In Figure 1-31 some details of the modern residential envelope are illustrated. Here the discrete envelope is clearly visible. Once the wooden frame is erected, including the plywood sheathing, the building paper and

1-31. The Modern Envelope (Residential Buildings).

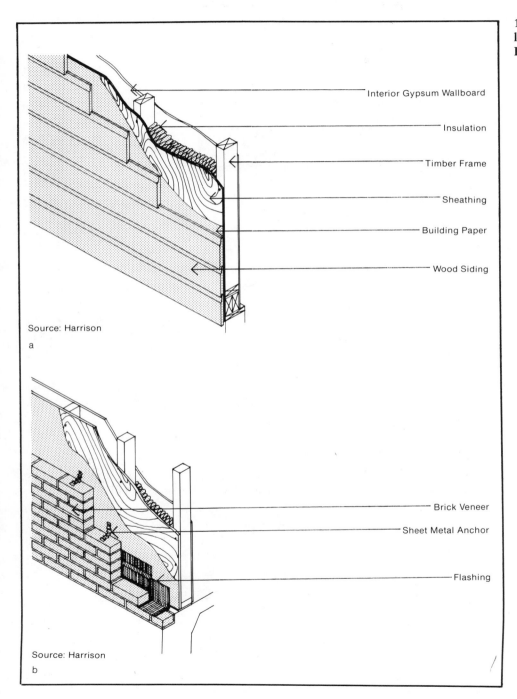

Interior Gypsum Wallboard

Insulation

Timber Frame

Sheathing

Building Paper

Wood Siding

Source: Harrison

a

Brick Veneer

Sheet Metal Anchor

Flashing

Source: Harrison

b

siding are applied. Insulation fills the cavities between the wood studs, giving the wall some semblance of thermal performance. Windows frame into this wall system much the same as they did in the medieval wood structure. Figure 1-31(b) shows the application of a brick veneer, which can give a building the appearance of being a brick load-bearing structure of the type popular in Georgian England.

As with major buildings, advances in glass, masonry, and insulation technologies have affected the evolution of the residential envelope. Plate glass has brought about a significant lightening of the wall surfaces. As the stone veneer has become more costly in masonry look-alike homes, the amount of glass used has increased to offset the expense of the veneer. An extreme example of this may be seen in Figure 1-32, which shows Philip Johnson's "Glass House." The masonry veneer has evolved in much the same way in small buildings as in large ones—it has become a thin "mask" anchored to the frame. The production of insulating materials in this century, especially glass fiber or mineral fiber blankets sized to fit between wood studs, has encouraged the development of this mask. Recent developments have made possible a combined sheathing and insulating material.

In summary, the envelope in residences has evolved similarly to that in larger buildings. Whether it was brick nogging between timber posts and lintels, or glass windows set in a masonry frame, the early envelope filled in where the frame left off. Since the advent of the balloon frame, plate glass, and insulating materials, the envelope has become a veneer of any material and appearance applied to the exterior of the frame, with ever larger expanses of glass. As with large structures, this veneer may reveal little about the makeup of the building within.

1-32. The Modern Envelope (Residential Buildings).

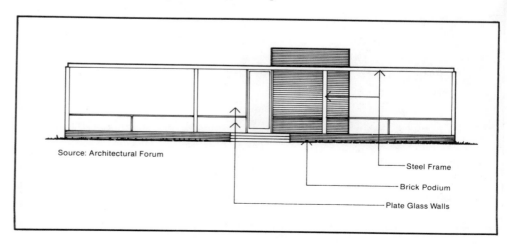

Source: Architectural Forum

Steel Frame

Brick Podium

Plate Glass Walls

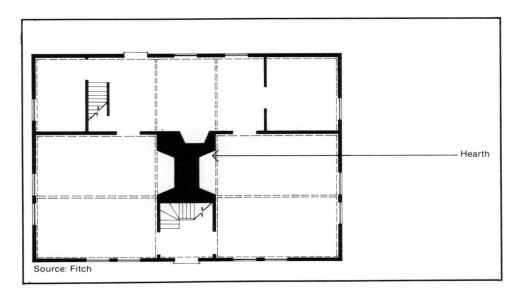

Hearth

Source: Fitch

1-33. Environmental Controls Prior to Industrialization (Residential Buildings).

The Machinery

Like major buildings, small structures have witnessed the advent of electrical and mechanical devices that have profoundly altered the physical and economic makeup of the building.

Colonial homes in America, such as the Baldwin House shown in Figure 1-33, had no active environmental control or sanitary features. Passive means, such as house shape and orientation and hearth location, were used to modify the extremes of the weather. The outdoor well and privy came closest to providing plumbing and sanitary accommodations for the colonists. Conditions on the continent were not much different, since the elaborate water supply system built by the monarchies did not come into being until the eighteenth century. Mechanical, electrical, and lighting advances would not occur until the 1800s.

In Figure 1-34, the state of the art in the application of the machinery to the home of 1869 is pictured, in Catherine Beecher's "American Woman's Home." (Beecher was a nineteenth-century domestic reformer who proposed this design as a prototype for the American house.) Here the impact of plumbing and ventilation can be seen. A spine containing the water supply and waste lines, as well as the exhaust vents and hot water heating lines, runs up through the center of the building. From this spine, radiators, water closets, sinks, and ventilation ducts branch off to serve particular spaces. As Banham points out, this is a thoroughly modern conception of the arrangement of machinery in a home.

1-34. Environmental Controls During Industrialization (Residential Buildings).

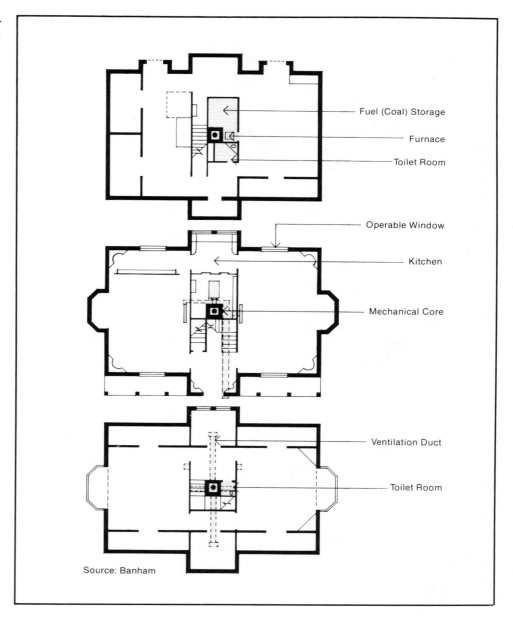

Fuel (Coal) Storage

Furnace

Toilet Room

Operable Window

Kitchen

Mechanical Core

Ventilation Duct

Toilet Room

Source: Banham

Figure 1-34 is remarkable also for its similarity to Figures 1-33 and 1-15. In all three, covering a wide range of building sizes and eras, the service spaces comprise the core around which the rest of the building spreads. In all three, the purpose was the same—the efficiency with which

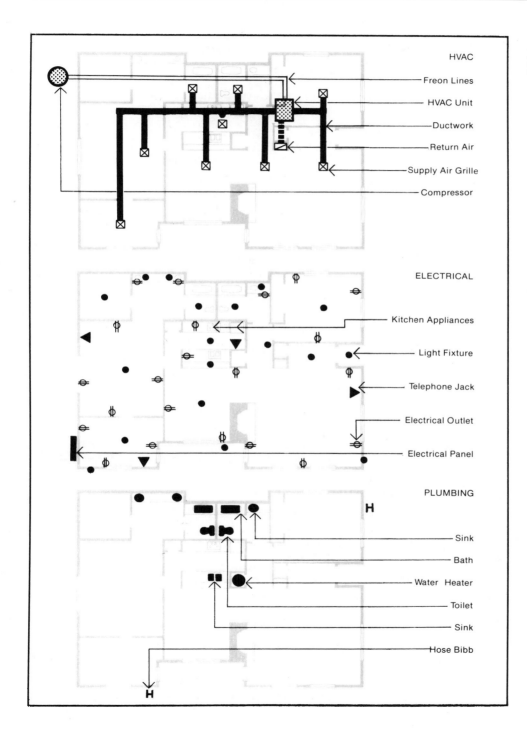

1-35. The Modern Machinery (Residential Buildings).

HVAC

Freon Lines

HVAC Unit

Ductwork

Return Air

Supply Air Grille

Compressor

ELECTRICAL

Kitchen Appliances

Light Fixture

Telephone Jack

Electrical Outlet

Electrical Panel

PLUMBING

Sink

Bath

Water Heater

Toilet

Sink

Hose Bibb

the environment-modifying devices might operate. Figures 1-34 and 1-15 show the extent to which the machinery has come to permeate buildings, whether small or large.

In recent residences, the culmination of Catherine Beecher's vision may be seen. Figure 1-35 shows the heating, ventilating, and air-conditioning, electrical, and plumbing distribution in a modern home. From the general core area in the upper center portion of the home, the HVAC branches out. The electrical branches off from the panel box at the lower left. Plumbing is all concentrated in the core area. The sheer amount of machinery-related devices is overwhelming. Older, grimier devices like the Franklin stove have been supplanted by a modern HVAC system. In addition, electrical power, which could not be foreseen in 1869, extends to all corners of the home. With the exception of the HVAC and the number of toilet rooms, this might be a typical home layout from the 1920s. Air conditioning was not introduced to houses to any extent until the 1950s, and then it was only in the form of packaged air-conditioning units that fit in windows. Heating was normally accomplished through radiators that circulated steam or hot water, or by electric heaters. It was not until the late 1960s that forced-air central HVAC systems, as shown in 1-35, would become commonplace.

In summation, the application of the machinery to houses has been a gradual process covering a period of two centuries. Nevertheless, three eras within this period are distinguishable. The first began about 1800 with the introduction of water closets and water supply to homes. Concurrently, gas light and hot water heating were introduced. The rudiments of environmental control were in place. The second era was ushered in by the domestication of the electrical supply in the late 1800s. The almost universal application of electricity in the West by the time of World War I significantly upgraded the interior environment. It provided a light source more consistent and luminous than gas and more flexible in its location. It spurred the development of household appliances, eliminating or reducing the grime of gas effluent and cooking devices. The kitchen could now be located integrally within the main body of the house. The final era began with the advent of air conditioning. The entire home environment could now be sealed and controlled.

The Infill
The history of the infill in residential construction is hardly as distinct as that in major buildings. For while large buildings saw the frame go from an all-encompassing masonry structure to a steel skeleton, small buildings still largely utilize load-bearing wall construction, which obviates the need

for an independent partitioning system. In spite of this, the infill is likely to be the single most significant cost component in residential work.

An account of the development of the infill materials was presented above. In Roman residences, the infill normally included little more than wall plastering and painting and some false plaster vaulting. As may be seen in Figure 1-36, the load-bearing walls at the House of the Surgeon in ancient Pompeii (fourth century B.C.) divided the spaces, just as they did later at the town house in Udine, Italy designed by Palladio in the 1500s, shown in Figure 1-37. In fact, substantial differences are not noted even in the twentieth century Robie House, illustrated in Figure 1-38. Spans are somewhat greater here and wall thicknesses less, but the basic system is the same. In a more recent home design from the 1970s (shown in Figure 1-39), the structure still divides the space, even if it is now wood instead of masonry.

In wood-frame houses of the Middle Ages, wattle-and-daub might be used for partitioning. However, many timber-framed houses, such as that shown in Figure 1-33, consisted simply of one compartment with no internal divisions. While pre-balloon frame buildings may have had some need for partitioning, they were not likely to have other decorative work, such as

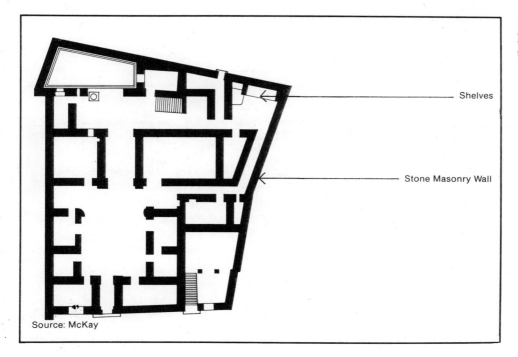

Shelves

Stone Masonry Wall

Source: McKay

1-36. The Infill Prior to Industrialization (Residential Buildings).

1-37. The Infill Prior to Industrialization (Residential Buildings).

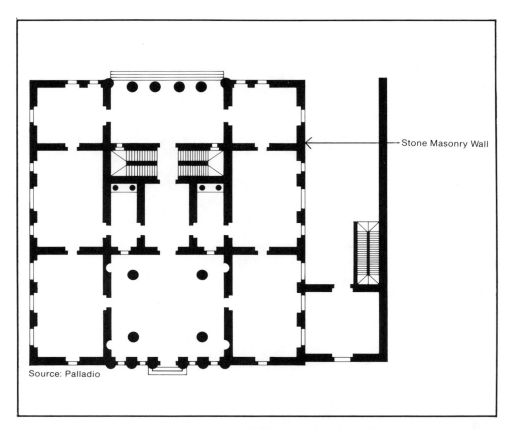

Stone Masonry Wall

Source: Palladio

1-38. The Infill During Industrialization (Residential Buildings).

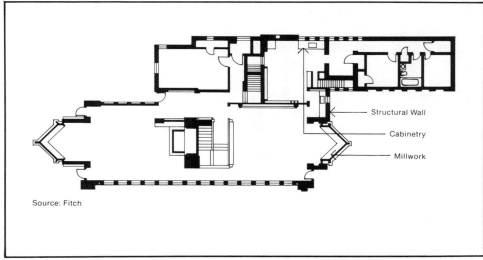

Structural Wall

Cabinetry

Millwork

Source: Fitch

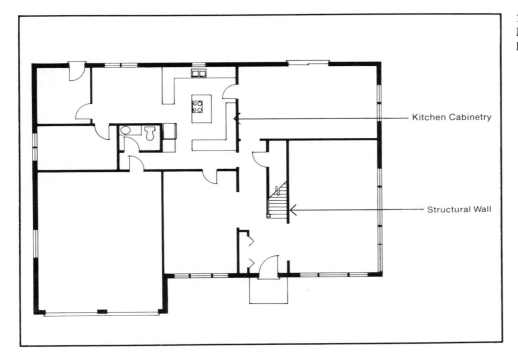

Kitchen Cabinetry

Structural Wall

1-39. The Infill in the Modern Era (Residential Buildings).

plaster vaulting. The heavy timber-framed house was thus more an anomaly than the rule in the history of house building.

In our age, the infill consists of gypsum wallboard attached to wood studs and floor joists, decorative treatments applied on top of the wallboard, and cabinetwork that goes into the home. Herein lie the reasons for the relative expense of the infill. Decorative treatments are likely to be more elaborate in the individual home than in the major building, where individual choices are not possible—wood bases replace vinyl, moldings and cornices are more prevalent, floors will more often be covered with tile or stone such as marble, and stairs will have decorative balustrades. Cabinetwork in residences is more widespread on a per-square-foot basis than in larger buildings—kitchens and bathrooms typically have built-in casework. Finally, the smaller rooms in residences translate into greater wall areas to be painted and treated, hence more cost.

In summary, infill technology in small-scale structures has differed from that in large buildings only in that, for the most part, it has not become a partitioning system. It has always been the decoration of the rough edges of construction and is now a thin facade behind which the frame, podium, and machinery hide.

Evolution of the Building Trades and Design Professions

Along with changes in building technologies came shifts in the organization of those responsible for putting the buildings together. Just as buildings changed from masonry to a plethora of specialized materials, a division of labor occurred in the design and construction trades that has served to further the fragmentation of the building process.

The seeds of the modern organization of a building project were sown in medieval times, with the growth of the trade guilds. Prior to that, Western civilization's major buildings were assembled using slave labor, as in Rome, or volunteer labor after the downfall of Roman civilization. In the Middle Ages, edifices were constructed largely by the masonry and carpentry crafts. Within the masonry guilds there existed a few levels of specialty crafts. Quarriers removed stone from the quarries, and either they or rough-hewers cut it into the basic shapes to be used in building. Freemasons were those skilled in the ornamental carving that graced the surfaces of the stones; sculptors grew out of this particular branch of the masonry guild. Finally, the setters put the stones in place, with the help of the navvies or laborers. Glaziers, plumbers, and other minor tradesmen were also needed to complete the building. Such was the trade organization in medieval England; the rest of Europe was not much different. Each trade essentially concentrated on one component part of the building, and this same arrangement has been expanded upon in modern times.

Today a multiplicity of specialized subcontractors erects a building under the direction of a general contractor or construction manager. Each subcontractor is likely to deal with one specific aspect of the building and employ members of one particular trade union in the field. Off-site preparation of materials may be done by members of another union or perhaps nonunion laborers. Figure 1-40 shows the organization of the work force at a major construction site in both the medieval period and today.

Not only construction labor has been apportioned. The design of projects has passed from the hands of the master mason to a set of design professionals with responsibilities for discrete portions of the building. Three branches of these professionals have developed, each supplemented by various consultants who focus on detailed aspects of the design.

As the leading professionals in design, architects establish the overall configuration of the building. After this initial stage, architects then concentrate on the envelope and the infill. Design of the podium and frame is passed

Comparison of Trades Involved in Major Building Projects—Middle Ages versus Twentieth Century

Building Component	Middle Ages*		Twentieth Century**	
	Trade	Tasks	Trade	Tasks
Podium	Setters	Set stones in foundations	Operating Engineers	Excavation
	Laborers	Excavation	Laborers	Pour concrete and strip forms
			Carpenters	Concrete formwork
			Ironworkers	Place steel reinforcing bar and stairs
			Elevator Mechanics	Elevator Installation
			Tile, Terrazzo, and Marble Setters	Ground level exterior paving
			Pile Drivers	Pile foundations
Frame	Carpenters	Wood Centering (formwork)	Ironworkers	Erect structural steel and place re-bar
	Quarriers	Quarry and cut stone	Carpenters	Concrete formwork
	Setters	Set stone	Cement Masons	Concrete finishes
	Freemasons	Carve stone	Laborers	Pour concrete and strip forms
Envelope	Glaziers	Glass manufacture and installation	Ironworkers	Curtain-wall frame and backup
	Plumbers	Lead roofing installation	Stone Masons	Cut and set stones
			Carpenters	Insulation installation
			Glaziers	Glass
			Bricklayers	Brick and concrete masonry

continued

Comparison of Trades Involved in Major Building Projects—Middle Ages versus Twentieth Century

Building Component	Middle Ages*		Twentieth Century**	
	Trade	Tasks	Trade	Tasks
Envelope			Roofers	Roofing, caulking, and waterproofing
			Plasterers	Plaster and stucco
			Lathers	Lath
Machinery			Electrical Linemen	Electrical
			Electrical Wiremen	Electrical
			Boilermakers	Plumbing equipment
			Plumbers	Piping, fixtures, and accessories
			Pipefitters	HVAC equipment and piping
			Sheet Metal Workers	Ductwork
			Sprinkler Fitters	Entire sprinkler system
Infill	Plasterers	Plastering	Ironworkers	Ornamental metal
			Stone Masons	Stone veneers
			Carpenters	Drywall, doors, hardware, and ceilings
			Bricklayers	Masonry partitions
			Millwrights	Paneling and casework
			Carpet and Linoleum Layers	Carpet and resilient flooring placement
			Tile, Terrazzo, and Marble Setters	Paving and ceramic tile

1-40. continued

Comparison of Trades Involved in Major Building Projects—Middle Ages versus Twentieth Century

Building Component	Middle Ages*		Twentieth Century**	
	Trade	Tasks	Trade	Tasks
General	Smiths	Fabrication of metal tools used in construction	Teamsters	Materials transport; deliveries and hauling
	Carriers	Materials transport		
	Navvies	General labor	Operating Engineers	On-site equipment
	Laborers	General labor	Laborers	Clean-up

Source: Knoop and Jones
** Source: Thomas F. Bellows and Wage Cost Bulletin *(July 1, 1983 through September 30, 1983) of Construction Employers' Association of Texas; includes only on-site trades*

1-40. continued

on to the second branch of design, the structural engineers. Finally, the design of the machinery is given to the mechanical/electrical/plumbing (MEP) engineers. Architects themselves have spun off a branch called interior designers, who often assume the responsibility to design the infill.

James Marston Fitch, in his book *American Building, The Historical Forces That Shaped It*, places the appearance of the engineer as a professional in the mid-1800s and notes it as a decisive point in the history of American building. The complexities of materials and engineering techniques had made the master mason obsolete. The resulting divorce of the architect from the engineer, however, has produced a division of labor that has not always worked to the advantage of the art of building.

In the course of the past thirty-five years, each of the five components of buildings has grown more independent than ever and has developed an economic life all its own. Separate design professionals, consultants, craftsmen, and code officials each focus on one component and are occasionally contracted with individually by clients. Design now consists of an architect devising an envelope and infill that will conceal the work of the structural and MEP engineers. Figure 1-41 summarizes the differences between the medieval and modern design team.

1-41.

Comparison of Design Professions Involved in Major Building Projects
Middle Ages versus Twentieth Century

Building Component	Middle Ages	Twentieth Century Design Professional	Twentieth Century Consultants
Podium	Master Mason	Structural Engineer	Elevator Consultant Soils Consultant
Frame	Master Mason	Structural Engineer	
Envelope	Master Mason	Architect	Structural Consultant Curtain-wall Consultant Masonry Consultant Window Washing Consultant
Machinery		Mechanical and Electrical Engineer	Acoustical Consultant Lighting Consultant
Infill	Master Mason	Architect Interior Designer	Acoustical Consultant Lighting Consultant

CHAPTER 2

Component Costs in the Present Age

An understanding of the technology and trade organization involved in the building process will in itself give some indication of the cost apportionment in buildings through the ages. Actual data is available, however, which may help establish more definitively the relative component costs through the years. The first step in establishing this cost progression analysis will be to determine relative component costs in the present day.

In Figure 2-1, the value of new construction in the United States over the past decade is summarized. Principal building types are delineated and subcategorized. It is interesting to note that residential construction accounted for over 60 percent of all new construction in the United States in this time period, and that single-family houses alone represented 43 percent of all construction. In nonresidential building types, office structures were predominant, accounting for over 10 percent of all new building value in the private sector.

In order to determine the component cost breakdowns for each category of building, information from recent editions of the two principal national cost estimating guides, Dodge and Means, was examined. The purpose was to establish a generalized component cost assessment that would apply to twentieth century buildings, both residential and nonresidential.

In establishing the cost analysis for nonresidential buildings, cost percentage breakdowns for the podium, frame, envelope, machinery, and infill for several office building types were identified from Dodge and Means. These are illustrated in Figure 2-2. Cost breakdowns across the different office building types are fairly consistent. It appears that frame costs account for less in the smaller buildings; other than that, no patterns are discernible. So-called "architectural" items—envelope and infill—as opposed to "engi-

2-1.

Value of New Private Construction in the U.S. Authorized by Building Permits: 1973 to 1982

Building Type	Amount (in $billions)	% of Subtotal	% of Total
Nonresidential			
Office	$ 71.7	25.6	10.1
Industrial	60.5	21.6	8.6
Stores and Mercantile	57.5	20.6	8.1
Institutional	20.0	7.1	2.8
Other	70.2	25.1	9.9
Subtotal	$279.9	100.0	39.5
Residential			
Single-family Residences	$304.4	71.3	43.1
Multifamily Residences	122.3	28.7	17.4
Subtotal	$426.7	100.0	60.5
Total	$706.6		100.0

Source: Statistical Abstract of the United States

neering" items, account for at most 45 percent of the total cost. The single most expensive component is the machinery, which points out the emphasis we Americans place on environmental comfort in the work place. Less money is spent on the envelope than on any other component, which is not surprising when one considers that most building owners consider the envelope as merely an enclosure for the interior environment and not, as architects would have it, a source for artistic embellishment.

In Figure 2-3, industrial building types are examined. Although cost percentages across the different types are fairly uniform for each component, some exceptions do stand out. Computer centers and food processing plants apparently require considerably more expense for the environmental machinery and both take it out of the frame. Both computer centers and refrigerated warehouses need more money for the infill—the former are characterized by work stations and the latter by built-in insulated rooms. Surprisingly, the machinery costs in refrigerated warehouses are relatively low. The high cost of the envelope in warehouses may reflect the growing tendency to locate them in settings where appearance is important.

Stores and mercantile buildings are analyzed in Figure 2-4. In this category significant differences in cost allocations are clearly recognizable. Restaurants spend a considerable amount of money on the infill. This is a clear reflection of surveys that show that next to cleanliness, ambience is the principal

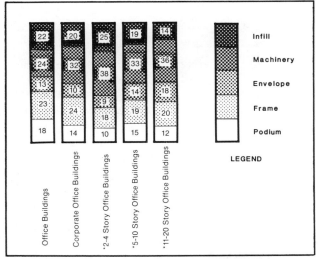

2-2. Component Cost Percentages in Office Buildings.

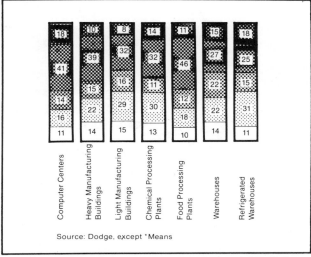

2-3. Component Cost Percentages in Industrial Buildings.

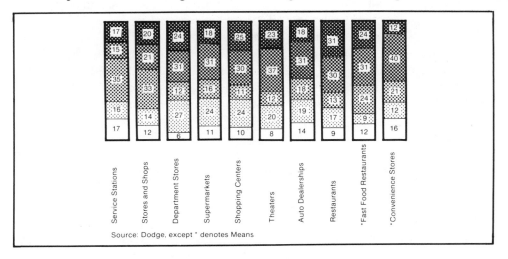

2-4. Component Cost Percentages in Stores and Mercantile Buildings.

marketing tool for eating establishments. Fast-food restaurants are also among the highest in terms of percentage of construction cost spent on the infill. Shopping centers and department stores also spend a great deal on the infill to entice customers into their confines, while service stations, supermarkets, auto dealerships, and convenience stores spend little. Apparently, the latter group consists of product-oriented enterprises in which extra pizzazz in the interior design does not greatly affect consumer choice, since they rely on their exteriors to convey their marketing message.

Understandably, service stations do not evidence a great deal of expenditure on the machinery, since relatively little of the floor area is environmentally controlled. It seems somewhat puzzling, however, that Dodge shows a relatively low figure for the machinery in stores and shops, while Means shows a high figure in convenience stores.

For small establishments where the potential customer is riding in an automobile, the building envelope is a major marketing tool, and this is reflected in the relative cost of this component. The consumer has made his choice by the time he drives into the parking lot. Service stations and stores and shops especially, but also fast-food restaurants and convenience stores invest a good deal of money in their billboards—the exterior walls. Shopping centers, department stores, and theaters have relatively little invested in the envelope, but this may be attributable to economies of scale due to their size and also to the fact that their sheer size is an attention-getter for highway travelers. Larger establishments are also more often the destination of a shopping trip, whereas small stores rely on passerby business.

Perhaps the most interesting building type in the stores and mercantile category is the fast-food restaurant. Proprietors of these establishments seemingly spend as little on the structure—podium and frame—as possible, provide adequate environmental comfort, and put most of their money into the envelope and the infill in order to attract and hold their customers. Nearly the opposite of the fast-food restaurant is the supermarket, where architectural embellishment, whether on the inside or outside, is not seen as such an important marketing tool.

Component cost percentages for institutional building types are presented in Figure 2-5. Some interesting observations may be made from this figure. Lower cost percentages for the envelope are generally seen in the educational building types, which would reinforce the notion that schools are places where image is not valued and that monumental building exteriors are not influential in parents' decision making regarding their children's education. On the other hand, public libraries and churches seem to value the envelope. This may be due to the fact that both of these building types must incorporate a degree of symbolism, hence ornamentation, into their exteriors. The public library is often a community's most prized civic building other than the city hall. Because of their interior heights, churches often have exterior walls of greater area in relation to the floor area than do most other building types.

Machinery costs are highest for laboratory buildings, many educational buildings, and hospitals and health centers. Special HVAC, electrical, and plumbing accommodations for the equipment in these building types account

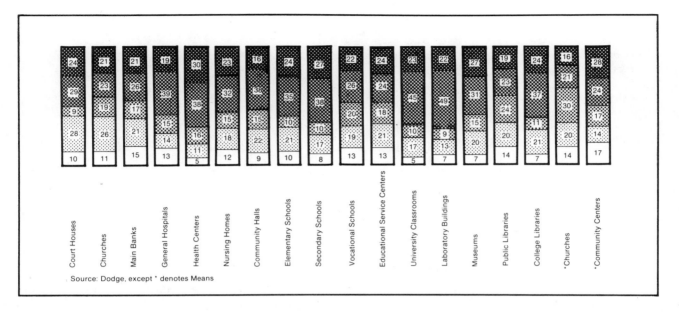

Source: Dodge, except * denotes Means

for this. Churches, on the other hand, are on the low end in money spent on the machinery, probably due to the occasional nature of their use.

Surprisingly, bank buildings rank about average in all component percentage costs. The need for such institutions to convey an image of strength and permanence, one might suppose, would be reflected in the envelope and infill costs being higher than average. Banks, however, seem to have more money allocated to the structure of the building. Perhaps most banks do convey that strong image by exposing that structure on which much of their money has been spent.

In Figure 2-6, miscellaneous building types are examined. Due to their dissimilarities, it is difficult to draw comparisons. Some things do stand out, however. The percentage of building costs spent on the infill in prisons is low, almost in the range of industrial buildings. The percentage spent on machinery is relatively high, however. Perhaps inmates are kept in sparsely finished quarters but enjoy a good degree of environmental comfort and sanitation.

The distribution of costs for country clubs is surprising. It appears that interior comfort is sacrificed for exterior image. Health and racquet clubs devote a good deal of money to facilities inside, while spending surprisingly little on the machinery.

Average component costs for each of the nonresidential building type categories in Figure 2-2 through 2-6 are shown in Figure 2-7. From this,

2-5. Component Cost Percentages in Institutional Buildings.

2-6 (*left*). Component Cost Percentages in Miscellaneous Nonresidential Buildings.

2-7 (*right*). Comparison of Average Component Cost Percentages in Nonresidential Building Types.

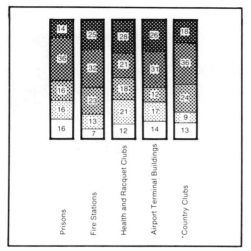

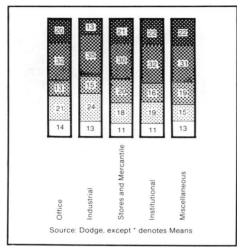

Source: Dodge, except * denotes Means

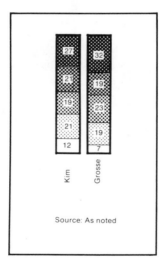

Source: As noted

2-8. Component Cost Percentages in Single-family Houses.

interesting insights may be gained regarding the degree of importance owners of different building types attach to each part of their buildings.

As the single most expensive component, the machinery will consume about one-third of the building budget, regardless of building type. Controlling the interior environment is the essence in the making of a building, and this applies equally to all nonresidential buildings. Depending on the complexity and purpose of the building, money spent beyond the machinery will be shifted between the structural and the architectural items.

In office and industrial structures, the podium and the frame consume more of the budgeted moneys at the expense of the envelope and infill. Offices generally have high podium costs associated with the conveying systems, and the high-rise frame represents more of a cost item than the frame in most other building types. Industrial edifices show a high percentage of cost expended on the structural components principally because so little (28 percent) is spent on the architectural features and not because of any complexity in the podium or frame itself.

On the other extreme are stores and mercantile buildings and institutional structures. These devote 30 percent or less to the structural components and about 40 percent to the architectural elements. While stores and mercantile buildings generally spend a good deal of money on the envelope, institutions put more into the infill. As noted before, this reflects the commercial aspect of stores and the use of the building facade as a merchandising tool, whereas institutions generally stress the services offered within the building. The percentage averages for miscellaneous building types approximate those for

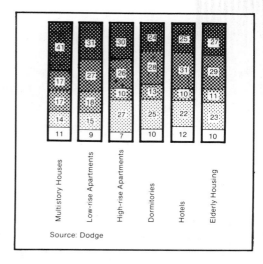

Source: Dodge

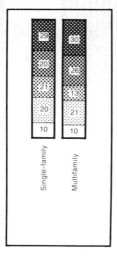

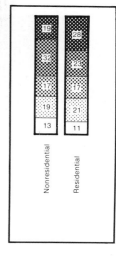

2-9 (*left*). Component Cost Percentages in Multi-family Housing.

2-10 (*center*). Comparison of Average Component Cost Percentages in Residential Building Types.

2-11 (*right*). Average Component Cost Percentages for Nonresidential and Residential Building Types (Weighted According to Building Permit Valuations).

stores and mercantile structures, but due to the diversity of building uses, no conclusions may be drawn.

Neither Means nor Dodge contain enough information to determine component cost breakdowns for single-family residences. To determine these percentages, construction inspection reports used by custom home builders in Houston, Texas were used. In Figure 2-8, the breakdowns for single-family residences are shown. It is interesting to note that, with minor variations, the percentages spent on each component are roughly equal. The infill, however, is the primary component.

In Figure 2-9, multifamily housing cost percentages were drawn from Dodge. In multistory houses and low-rise apartments, considerably less of the building budget is consumed by the frame than in the other building types. This is probably due to the lower cost of wood-frame construction as opposed to high-rise concrete or steel construction.

Low-rise buildings put more into the envelope than their high-rise counterparts. A possible explanation for this is that stylistic variety is more often found in low-rise housing, since it must appeal to people at a pedestrian level, and individual unit identity is desirable. In high-rise structures, pedestrian approach is on the inside of the building—thus the envelope may be made of standardized pieces.

With the exception of multistory houses, the cost percentages for the machinery and infill are nearly uniform across all building types in Figure 2-9. Perhaps the likelihood that multistory houses are owned, rather than rented, explains the greater expense put out for interior architectural treatments, as opposed to machinery. However, single-family residences, as shown in Figure 2-8, have the same condition of ownership, yet the infill cost is not nearly so high.

Average component cost percentages for single-family houses and multi-family residences are compared in Figure 2-10. This figure might be said to compare the condition of home ownership versus home rental. In single-family homes—those likely to be owner-occupied—it appears that the desire for individual identity and status is reflected in the higher percentage of cost allocated to the envelope. In multifamily housing, the percentage expended on the machinery is higher because the dwelling units are smaller, and there are likely to be more toilets, sinks, and baths per total building square foot than in detached homes. Since there are more kitchens per total building square foot, and hence more cabinetwork, the infill is also more costly in the multifamily category. Infill expense in the single-family home nearly equals that in multifamily dwellings, however, which would suggest that home owners are likely to spend more money on finish materials than are apartment builders.

Having established component cost percentage breakdowns for each of the building types shown in Figure 2-1, an average for nonresidential and residential buildings may now be determined. For nonresidential buildings, the averages illustrated in Figure 2-7 were weighted according to the percent of nonresidential construction value each type represented. For residences, the averages from Figure 2-10 were weighted by the percent of residential construction value for each type. The resulting averages are shown in Figure 2-11 and will be used in historical comparisons that generalize component costs in the present era.

CHAPTER 3

The Historical Evolution of Component Costs

In order to establish the cost percentage progression of each building component from Roman times up to the present, cost information for similar building types from different eras will be compared. Major buildings will be examined first. In Knoop and Jones and in Salzman, detailed financial accounts for certain medieval building projects are presented. These were used as the basis of the cost estimates shown below for the buildings from the Middle Ages.

Figure 3-1 shows the percentages of total building cost spent on each component in what may be termed civic buildings. Percentages from the Ipswich Shire House, a medieval English structure, are compared with those from two recent American buildings. One is immediately struck by the impact of the machinery on the economics of modern buildings. In 1442, active alteration of the interior environment was insignificant; by 1979 it was a necessity.

Also evident in Figure 3-1 is the shift of a good deal of money allocated to frame construction in the Middle Ages to the infill today. Curiously, as the frame dropped in relative importance, so did the envelope. This does not reflect a decrease in material resources dedicated to the envelope as much as it illustrates the increased costs of the machinery and infill. Most major buildings of the medieval period actually had relatively less invested in the envelope than the wood-framed Ipswich building.

In Figure 3-2 the nature of masonry building in the Middle Ages is more apparent. The frame was clearly the dominant cost component in the construction of Eton College in fifteenth-century England. Machinery expense in 1450 was negligible. In the cost/percentage breakdowns for similar college buildings in recent years, the predominance of the machinery and infill is

3-1 (*left*). Historical Comparison—Civic Buildings.

3-2 (*center*). Historical Comparison—Educational Buildings.

3-3 (*right*). Historical Comparison—Church Buildings.

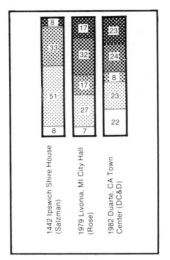

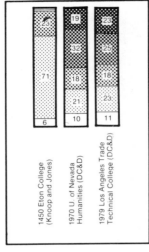

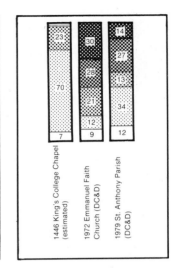

3-4 (*left*). Historical Comparison—Commercial Buildings.

3-5 (*right*). Historical Comparison—Miscellaneous Building Types.

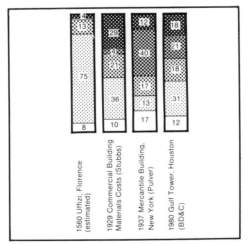

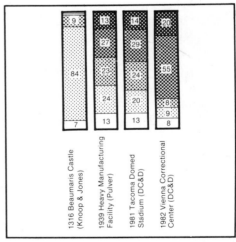

again quite clear. About half the construction budget is now consumed by these two components.

Church buildings from fifteenth-century England and twentieth-century America are compared in Figure 3-3. Once again, the shift of resources into the machinery and infill, and away from the frame, is remarkably clear.

In Figure 3-4, the estimated cost breakdown for the Uffizi in Florence, derived from historical accounts of the building itself and of the structure of the continental building industry (from Goldthwaite and others), is presented. This is the first of this building type—commercial offices—and according to Pevsner it gave the name to office buildings. Cost breakdowns for two early twentieth-century office buildings, as drawn from cost estimating literature, are also illustrated. Finally, the Gulf Tower, from 1980, is compared with the rest. This figure clearly illustrates just how recently the machinery has made its mark on buildings. It also suggests that, like many technologies, air conditioning was more expensive in its novel stages, circa 1937, than it has been since. Refinements and mass production have lowered its unit cost.

Figure 3-4 also suggests that envelope and infill cost percentages may have bulged in the period after the steel frame supplanted masonry and before the full use of environmental controls was exploited. As frame expense waned, buildings continued to be designed in the prevailing historical revival styles. But while technological advances cut frame costs, they drove up the expense of the hand-worked stone crafts that accounted for much of the envelope and infill elements. This was due to the fact that industrialization essentially shifted all functions except final assembly from the field to the factory. In the shop, the use of machines and assembly-line techniques greatly increased the net output of each worker and served to decrease labor cost as a percentage of material unit cost. Thus, labor-intensive trades that did not or could not take advantage of the machine became more costly, before experiencing decline.

Miscellaneous building types are compared in Figure 3-5. These range in time period from fourteenth-century England to the twentieth-century United States. Again, a large proportion of the building budget is now taken by the machinery and infill, whereas these components had been nearly nonexistent at Beaumaris Castle. More evidence is available here to support the notion of the pre-air-conditioning bulge in envelope cost percentages. The 23 percent for the heavy manufacturing building in 1937 is 8 percent higher than the average for modern industrial structures.

Interpolations and extrapolations from this information, supplemented by additional data, yield a cost percentage breakdown that traces each component in major buildings from the time of the Pax Romana (first century A.D.) to the present. This is illustrated in Figure 3-6. Cost percentages for each component in the fifteenth century are approximated using Figures 3-1 through 3-5 as a basis. Percentages for the twentieth century are the averages for nonresidential building types from Figure 2-11. Actual interpolations and extrapolations are based on the historical accounts of each component technology outlined in chapter 1 and miscellaneous data that will be examined in more detail below.

Among the more interesting aspects highlighted in Figure 3-6 is the

3-6. Historical Progression of Component Cost Percentages in Nonresidential Building Types—0 to 1984 A.D. by Century.

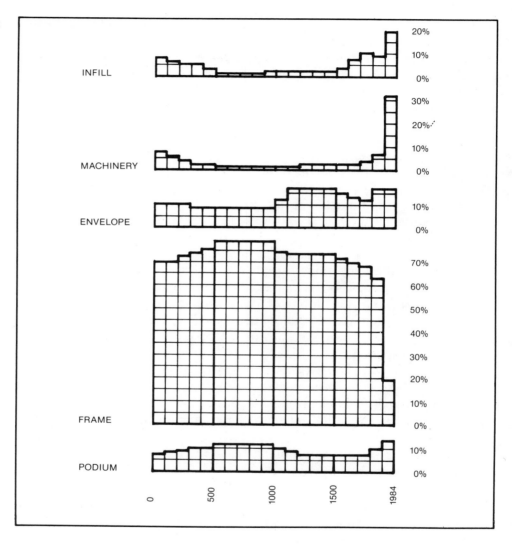

relative stability of the podium and the envelope as economic components of buildings. Perhaps it is not surprising that the podium consistently represents such a relatively small expense. Building owners cannot be expected to spend more than the minimum for what is essentially a hidden and perfunctory part of the structure. With all of the changes in architectural style through the centuries, however, it seems unusual that the envelope cost would not have fluctuated more.

Probably the main factor accounting for the rise in the percentage of money spent on the frame from Roman times to the Middle Ages is the fact that there was a decline in absolute terms in the resource cost of major buildings during the Dark Ages. While building mass was not minimized, structural knowledge that could have helped decrease mass was nowhere near the level of sophistication of the Gothic era. Labor was minimized, transportation of materials was kept to a minimum, and finishing and specialized crafts were sparsely used. After the Roman Empire's water supply system disintegrated, machinery cost in buildings was virtually nonexistent for centuries. Buildings of this period were little more than basic shelter and fortification.

With the onset of commercial expansion in Europe in the medieval era, an increase in resource cost in absolute terms began that has continued to this day. In the twentieth century, much of this resource cost is not the capital improvement represented by the structure itself. Some of the cost is indirect, namely operating expenses incurred largely to maintain the interior environment. Such expenses are not dealt with here.

Money once spent on the frame of the building is now put toward the control of the interior environment and to the interior partitions and finishes. The latter are merely substitutes for the frame in functions once performed by the carving of masonry surfaces. The former have revolutionized buildings to the extent that the machinery virtually *is* the building. Interestingly, architects want little to do with the most important cost component in buildings. The design of the machinery is left to engineers.

In order to get a better understanding of the period of monumental changes over the last century-and-a-half (of which Figure 3-6 gives only a glimpse), Figure 3-7 presents a decennial breakdown since 1800. Until about 1850 the medieval building type was the mainstay of major construction efforts. With the introduction of steel, frame costs began a rapid decline. Prevailing tastes, however, still coerced architects into the design of buildings that utilized the medieval decorative method. As a result, the envelope's cost percentage rose as sculpted masonry continued to be used as its building material.

It was merely a matter of time before the economics of the situation effected change, and advances in the technology of the machinery provided the incentive. As environmental controls became more essential, building owners sought to eliminate nonessentials. The elaborate envelope, medieval in design and fabrication, was the victim.

Construction cost estimating literature makes apparent the meteoric rise in the investment that the machinery represents. Data from the nineteenth and early twentieth centuries reflects the relative insignificance of the machinery

3-7. Historical Progression of Component Cost Percentages in Nonresidential Building Types—1800 to 1984 by Decade.

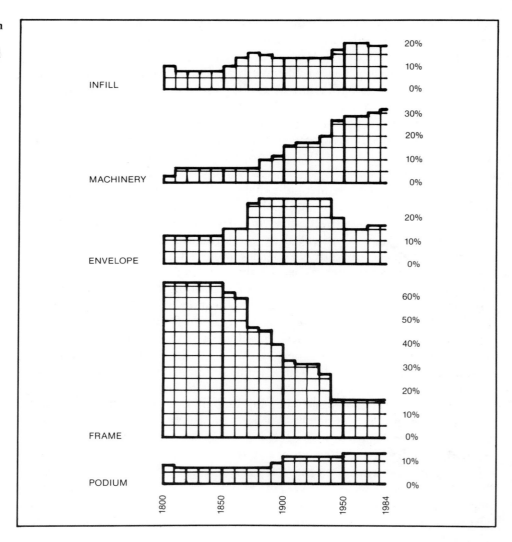

of that period. Robert Waters' *The Builder's Guide* of 1851 contained prices for carpenters, bricklayers, and other such tradesmen, but none for any mechanical or electrical trades. In Hodgson's *The Builder's Guide and Estimator's Price Book* from 1902, only plumbing and heating costs for cast iron boilers are noted. *Estimating*, published by Edward Nichols in 1913, devotes a mere eight of fifty-eight pages to heating, plumbing, gasfitting, and electric work. Nichols put the expense of providing heating devices in a building at 6 to 12 percent of the total cost. Gasfitting could be 3 to 5 percent, and plumbing up to 10 percent. Altogether, machinery costs, according to Nichols, might amount to 19 to 32 percent of the total building cost.

By 1931 Frank Edwin Barnes's third edition of *Estimating Building Costs* devoted 20 percent of its pages to mechanical and electrical construction. Barnes's book noted that electrical wiring would consume one to one-and-a-half percent of the building budget. He showed no costs for forced-air HVAC; only radiation heating is included in the mechanical work. In 1958, *Engineering News-Record* was estimating that HVAC, plumbing, and electrical work would take 34.8 percent of the construction budget for a large (242,000 square feet) office building. In 1983's *Square Foot Estimating* published by R.S. Means, 30 percent of the pages relate to machinery costs. A summary of the extent of cost-estimating literature devoted to the machinery is presented in Figure 3-8.

Another source for tracing the growth of machinery costs is Siegfried Giedion's 1948 book, *Mechanization Takes Command*. The author calls the era between the World Wars the "Time of Full Mechanization," during which electric appliances and the modern kitchen and bathroom permanently made their imprint upon design. Trevor Williams corroborates this in his

The Importance of the Machinery in Cost Estimating Literature   3-8.

Year	Author	Total Pages	Pages Re Machinery	% Pages Re Machinery
1851	Waters		0	0
1856	Wilson		0	0
1882	Hodgson	213	1	1/2
1913	Nichols	58	8	14
1922	Arthur	807	113	14
1928	Arthur	163	8	5
1929	Dingman	150	5	3
1931	Barnes	554	110	20
1940	Pulver	429	74	17
1942	Roberts	224	38	17
1950	Underwood	295	76	26
1953	Peurifoy	282	23	8
1957	Dallavia	177	0	0
1958	Peurifoy	400	78	20
1978	Van Orman	243	30	12
1983	Means	214	64	30
1984	Dodge	115	26	23

1982 work, *A Short History of Twentieth Century Technology.*

Nearly as significant as the growth of the machinery since 1800 was that of the infill. From 1800 to 1983, sheer quantity was substituted for ornamentation in the infill. Similar to the envelope, the infill experienced an adjustment period when steel became the frame material. Mass-produced, field-assembled, lightweight standard elements came to comprise an infill that was largely a partitioning system.

Residential buildings present a somewhat different picture from large structures. In Figure 3-9, two buildings of wood frame construction from medieval England are compared with nineteenth- and twentieth-century residences. Although the impact of the machinery and the infill is once again evident, an equally interesting story is told in what happens with the envelope. In the Middle Ages, this component constituted a significant portion of the residential building. As industrialization affected housing construction in the 1800s, the frame's value decreased, the infill became significant, and the envelope became predominant. By the 1920s, owners of buildings real-located money to the machinery that had once been set aside for the envelope. Finally, by the time of the Second World War, the envelope had become the least significant economic constituent of the building, with the exception of the podium.

Since 1940, the distribution of money for each building component has remained fairly consistent. The infill is now the predominant component in residential construction, but the machinery is not far behind. Figure 3-10 compares the percentage expended on the machinery in the two medieval cases shown in Figure 3-9 with further examples of twentieth-century American homes. Even in houses with no air conditioning, the machinery is a substantial expense.

Interpolations and extrapolations once again yield a cost/percentage breakdown that traces each component cost from the time of the Roman Empire to 1984. This is illustrated in Figure 3-11. Figures in the fifteenth century are approximated from Figure 3-9, and cost percentages in the twentieth century are taken from Figure 2-11.

Unlike nonresidential construction, the residential envelope is not as uniform through the centuries. This reflects the shift from Roman masonry domestic construction to the heavy timber framing common across northern Europe in the medieval era with the corresponding increase in envelope costs. One could make a case that the envelope should not show such an increase, since domestic construction in the Mediterranean region remained masonry in nature. But timber-framed residential buildings more closely led to the typical home construction used in the United States today, where

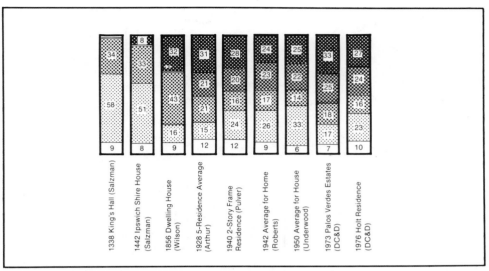

3-9. Historical Comparison—Residences.

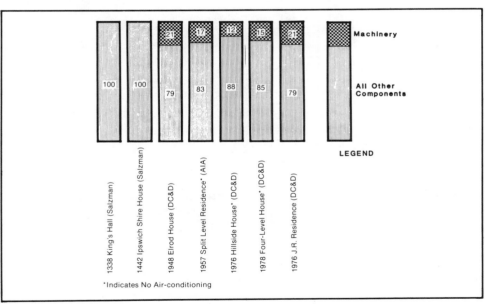

3-10. Historical Comparison—The Machinery in Residences.

the bulk of Western residential construction is conducted.

With the exception noted above, general trends in residences have been similar to those in large buildings. Differences in degree exist, however. The frame in the home has rarely accounted for such a large proportion of the total cost as in nonresidential buildings. Assuming that Roman masonry

3-11. Historical Progression of Component Cost Percentages in Residential Building Types—0 to 1984 A.D. by Century.

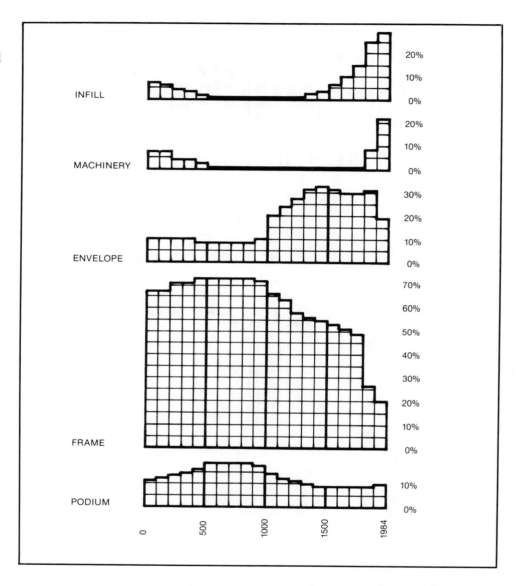

construction was universal in the West, the frame in homes was of the same proportion in large structures only until the demise of the Empire. Without the Imperial infrastructure or slave labor, Westerners turned to the more easily worked and transported timber for small buildings.

While the machinery may have been present to a minor degree in some major structures of the Middle Ages, it disappeared from domestic buildings

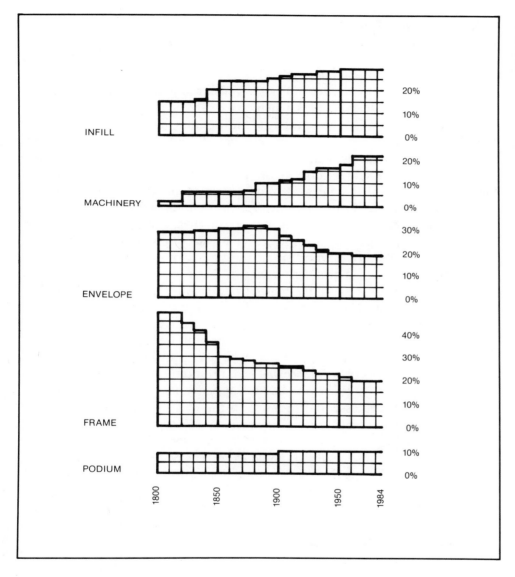

3-12. Historical Progression of Component Cost Percentages in Residential Building Types—1800 to 1984 by Decade.

altogether when the Empire's water supply system disintegrated. It would not reappear in residences until late in the eighteenth century. Mainly because of the large relative expense of the infill, machinery cost in homes today is not of the magnitude as in nonresidential buildings.

Although Imperial Romans made liberal use of some infill devices in homes, the features that make the infill so relatively expensive in today's residences were not used much in classical times. Wall surfaces and plaster vaulted ceilings in Rome could be quite ornate. Paintings were elaborate, and decorative cornices and friezes in plaster were not uncommon. But a major portion of the infill cost today is in the built-in cabinetwork in kitchens and bathrooms. Roman homes, according to McKay, might only have had some shelves in the kitchen.

As with nonresidential buildings, the podium has historically been a fairly uniform cost in residences. It has also been the least expensive component.

Figure 3-12 presents a decennial breakdown of component costs since 1800. This gives a closer look at the period of industrialization. As with large structures, the medieval building method prevailed until the mid-1800s. The balloon frame not only revolutionized home construction by significantly decreasing the frame cost of timber buildings—it made the masonry frame for residential construction an anachronism in North America. It made the building of a wood frame with a masonry veneer cheaper than constructing masonry load-bearing walls.

Similar to major buildings, the envelope on residences increased in relative cost as the frame decreased. The Victorian and historical revival styles of the 1800s emphasized elaborate, hand-crafted decoration. As industrialization proceeded, however, such elaboration became uneconomical.

Advances in the technology of the machinery provided the same impetus to cut envelope costs in residences as it did in nonresidential structures. By 1940, this economic "correction" had taken place. The control of the interior environment is more essential today than the decoration of the facade.

With all building types, both residential and nonresidential, it is interesting to note that some components seem to have cost/percentage limits above which owners are not generally disposed to go. The podium is most noticeable, falling at about 10 percent of the total building cost. With some exceptions, the envelope also appears to have been limited to about 15 percent of the construction cost in major buildings and slightly higher in dwellings. Fifteen percent is not a great deal when one considers that it is here that architectural embellishment is usually concentrated. The historical fluctuations are centered in the frame, machinery, and infill.

CHAPTER 4

Trends in Component Costs

In order to discern any trends in component cost proportioning, the Modern period of building was examined more closely. Cost information since World War II was looked at, such information having become increasingly available in periodical and serial publications since the 1960s. The predominant building type in each of the nonresidential and residential categories was selected as the subject for examination. These are office buildings and single-family residences. Small office structures as well as large ones were investigated. Figures 4-1 through 4-10 give a detailed view of the relative component costs in a sample of these office buildings.

Building size apparently has little to do with the percentage breakdown of the relative costs of the building components. Median percentages spent on the structural components (podium and frame) are only slightly higher in office buildings over 100,000 square feet in area than in those from 40,000 to 99,999 square feet. The envelope, and to a lesser degree the machinery, decrease in cost on a relative basis in the larger buildings. Overall, however, such variations are not very significant.

In spite of the relative stability of costs in recent decades that the figures illustrate, some trends are discernable. The historical decline of the percentage of cost consumed by the frame is continuing; the climb in the importance of the machinery is not abating; and the envelope is experiencing a revival of sorts. Trends in the podium and infill are less obvious.

A comparison in Figures 4-1 and 4-2 of the median cost percentages over the first half of the sample as opposed to the second half reveals that podium costs in smaller buildings have increased since 1959. In office buildings over 100,000 square feet, however, podium costs have declined slightly. Historically the most consistent cost element in buildings, the podium has

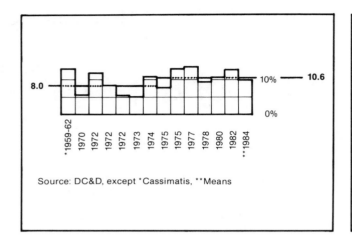

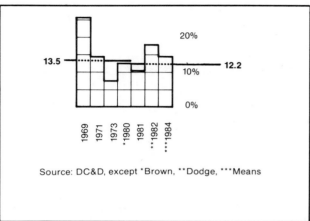

4-1 (*left*). **Recent Cost History of the Podium— Small Office Buildings (40,000–99,999 square feet).**

4-2 (*right*). **Recent Cost History of the Podium— Large Office Buildings (100,000+ square feet).**

seen few advances recently, save for improvements in elevators due to the application of solid state electronics. These advances have generally not borne a substantial enough price tag to affect overall cost distributions. Except for further innovations with respect to electronics in elevator controls, this situation is unlikely to change in the near future.

In both large and small office buildings, the percentage of cost expended on the frame was generally higher over the first half of the sample than over the latter half. This can be seen in Figures 4-3 and 4-4. Since the frame usually has no aesthetic role, structural engineers have sought to minimize the mass and the amount of field labor required in its erection. Improvements in engineering methods, composite steel and concrete construction, advances in fabrication and erection tools, precasting of concrete, and so forth have all contributed to the wide variety of choices available for frame construction. Steel and concrete today are generally competitive in price for use in buildings up to forty stories high.

In attempting to forecast the direction of frame costs, it is imperative to understand the factors that have driven these costs down in the past century. Not only is there an incentive in minimizing the weight of the frame materials themselves; the same incentive works on the envelope and the infill as well. As the overall mass of the building continues to decrease, so will that of the frame, and materials costs will therefore decline. Unlike the envelope and the infill, savings due to smaller materials costs are not likely to be offset by increased expenditures for architectural embellishment.

In contrast with the frame, relative envelope expenditures appear to be rising. Both large and small office buildings show an increase in the second half of the samples shown in Figures 4-5 and 4-6. One reason for this

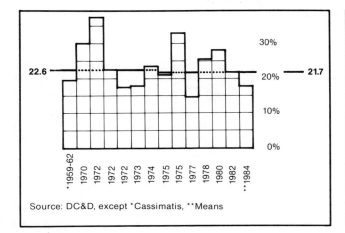

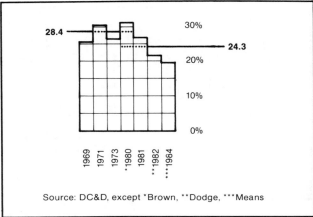

increase with respect to the median is that the cost of metal and glass curtain-wall systems has risen in recent years. Another factor driving up envelope costs has been the worldwide soaring of energy costs, which puts a premium on highly insulative exterior walls. Reflective glass, requiring high initial expense but possessing energy-saving qualities, is another innovation that has helped to drive up envelope cost.

At the same time that energy efficiency is demanded, reduced weight is demanded as well. Building designers have been learning the same things that automobile manufacturers have—the benefits derived from the use of high-strength steels for fasteners, aluminum and plastics for exterior panels, and reduced thicknesses for glass. Thus the envelope has become a thin weathertight veneer backed up by a bulky yet lightweight insulating material. Insulating materials have also gotten lighter, thinner, and more thermally efficient. As they have, the use of thin slabs of stone masonry anchored to steel backup systems has become feasible economically as well as technically. This has eliminated the previous drawback to stone—the great mass necessary to achieve satisfactory thermal properties.

The push for a lighter and more insulative envelope will continue— witness the increasing popularity of "exterior insulation" systems. These consist of expanded polystyrene insulation boards (Styrofoam) attached to light-gauge metal framing. The insulation is then covered with lath and a plasterlike material.

Another recent trend that is likely to continue, if not expand, is the architectural embellishment of the envelope. Since flexibility in automated production processes is on the rise, variations of standard building parts are now available without significantly affecting cost. Architects are now

4-3 (*left*). **Recent Cost History of the Frame— Small Office Buildings.**

4-4 (*right*). **Recent Cost History of the Frame— Large Office Buildings.**

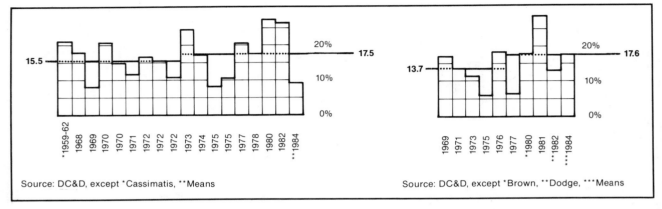

Source: DC&D, except *Cassimatis, **Means

Source: DC&D, except *Brown, **Dodge, ***Means

4-5 (*left*). **Recent Cost History of the Envelope— Small Office Buildings.**

4-6 (*right*). **Recent Cost History of the Envelope— Large Office Buildings.**

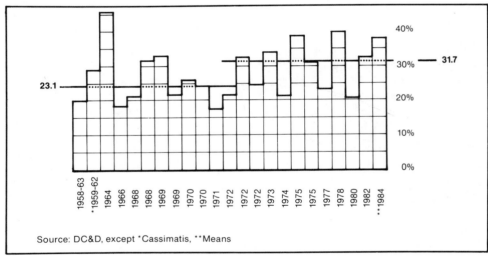

Source: DC&D, except *Cassimatis, **Means

4-7. Recent Cost History of the Machinery—Small Office Buildings.

also rediscovering the wide variety of ornament that has been available for years in standard industrialized product lines. These developments are meshing with the demand by clients for distinctive designs.

It appears that the one building component for which office building owners will continue to spend more money is the machinery. Figures 4-7 and 4-8 both show a substantial increase in relative machinery costs over the second half of the sample as compared with the first half. This is reflective of several trends of the past decade. The demand for energy-efficient buildings affected the machinery more profoundly than the envelope. Not only have HVAC, lighting, and electrical systems been designed to work more efficiently, electronics have been applied to these systems to coordinate them more

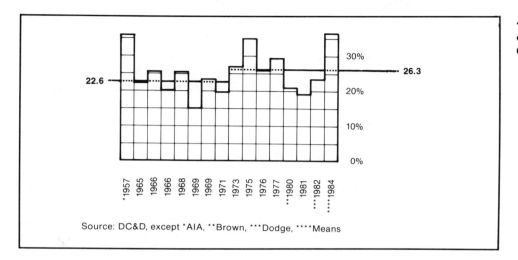

4-8. Recent Cost History of the Machinery—Large Office Buildings.

Source: DC&D, except *AIA, **Brown, ***Dodge, ****Means

effectively with exterior conditions. In recent years, life-safety requirements and computers have also increased machinery expense. An emphasis on fire protection in buildings, especially high-rise buildings, has grown. In many building types, codes now commonly require life-safety alarm systems, fire-extinguishing systems (sprinklers) and mechanical systems to exhaust smoke in the event of fire. The application of computers and electronics to the workplace has also increased demands on the HVAC system and created the need for sprinkler systems that do not use water. Halon and other gas extinguishing systems are becoming more commonplace for certain spaces in buildings.

Advances in electronics will also affect buildings. The effect on the overall composition of structures, however, should be minor. Current trends in the automobile industry give a good indication of why this is so. Womack and Jones have pointed out that the basic character of the modern automobile is not likely to change. Advanced technologies are being applied to the design and manufacture of cars and their adaptation to energy, safety and other standards. Likewise, the main impact of electronics on buildings will be in the speed and efficiency of their design and manufacture and in the efficiency of the building's operation. All components except perhaps the podium should feel this impact equally.

Not only can each of these trends be expected to continue, they are likely to accelerate in coming years. The machinery is the most necessary building component, and its relative cost has and will continue to reflect this.

Definite directions concerning the infill are more difficult to identify than in the other components. In Figure 4-9, median costs over the second half of the sample slightly exceed those in the first half. As may be seen in Figure 4-10, just the opposite trend occurred in larger office buildings.

Recent developments in infill materials, such as improved adhesives and machine tool technologies, have led to a greater variety and better quality of available veneer finishes. These veneers are thinner and lighter than ever, cutting costs for materials as well as for labor and handling.

Due to its nonutilitarian nature, the infill is a prime target for cost-cutting measures. The open plan was the architectural result of these economics. A recurrent vexation for architects is the building owner who in effect disrates portions of their work in order to bring the building within budget. In light of this, it is surprising that more designers have not sought instead to omit infill materials and upgrade the finish quality of the pieces of the utilitarian parts of the building.

It appears that the infill is not likely to grow in cost importance in buildings in the near future. Although the infill that remains may be covered with higher quality veneers, it will continue to decrease in material and weight. Moreover, factory prefabrication is readily adaptable to the infill, with further cost savings possible. And partitioning is often accomplished with furniture systems rather than with "hard" construction-cost items.

The infill is also affected by the machinery. As fire sprinkler systems are included in more buildings, fire separation requirements are reduced. Since such code-mandated separations are normally achieved through the use of partitioning or the like, the overall infill will decrease.

In Figure 4-11, component costs shown in Figure 3-7 are projected to the year 2000. In approximately 1950, an era of relatively stable component expenditure breakdowns was entered. This period continues today, although changes have occurred in recent years. The machinery and envelope are likely to take larger shares of the building construction budget. This will happen at the expense of the frame and the infill. Compared to the tremendous transformations in the century prior to 1950, however, these coming redistributions will probably seem insignificant.

Relative component costs in a sample of single-family residences are illustrated in Figures 4-12 through 4-16. While the historical decline of the percentage of cost consumed by the frame and the recent increase in machinery cost percentages applies to residences as well as office buildings, trends in the envelope and infill are somewhat reversed. The podium, as usual, remains at about 10 percent of the total building cost.

In Figure 4-12, the median percentage of podium expense remained

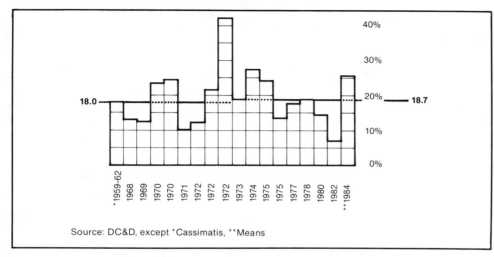

4-9. Recent Cost History of the Infill—Small Office Buildings.

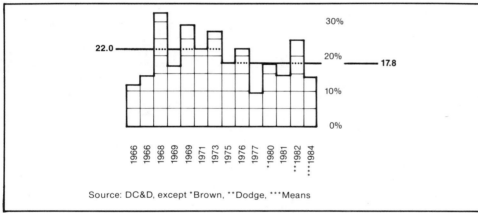

4-10. Recent Cost History of the Infill—Large Office Buildings.

fairly constant from the first half of the sample through the second half. In view of the nature and history of this component, as was shown in Figure 3-9, the proportion of podium cost is not likely to rise in the future above the 10 percent level.

Median frame costs were considerably less over the latter half of the samples than over the first half, as shown in Figure 4-13. There may be several reasons for this. First, the wood frame itself is still getting lighter. It is now common for the studs to be centered twenty-four inches apart, instead of the former standard of sixteen inches. Sheathing and the like have also undergone a metamorphosis. As late as the 1950s, sheathing and sub-flooring were commonly done with wood planks laid horizontally or

4-11. Projection to 2000 A.D. of Component Cost Percentages in Nonresidential Building Types.

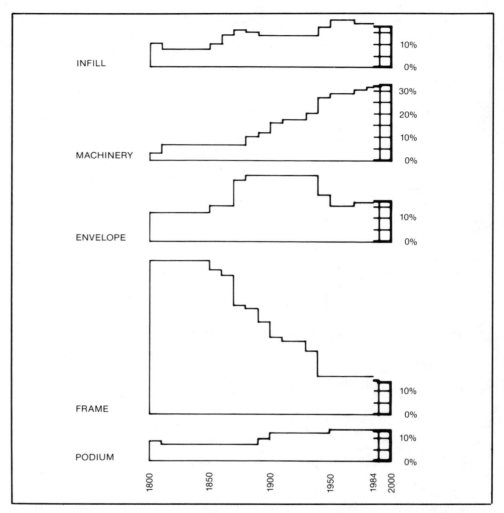

4-12. Recent Cost History of the Podium—Single-family Residences.

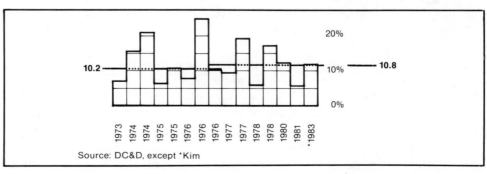

Source: DC&D, except *Kim

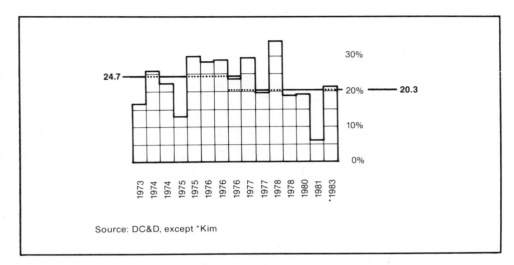

4-13. Recent Cost History of the Frame—Single-family Residences.

Source: DC&D, except *Kim

diagonally across studs and joists. Then plywood replaced the planks and considerably cut down the field labor involved. Plywood, in turn, is being replaced on walls more and more often with composition board or with insulative, lightweight sheathing. This further reduces the complexities of handling and field labor. Another innovation that has cut frame costs is the displacement of masonry chimneys with metal ones that can be placed in wood framing, then covered with masonry veneer. This has taken the cost of the brick fireplace out of the frame and made it an element of the infill.

Although the pressure to reduce frame costs will continue, one must wonder if the lower bounds have not already been reached. As long as wood stud construction remains the mainstay of single-family residence construction, it is not easy to see how frame costs can be reduced much further. This seems to be recognized in the marketplace, where homebuilders now must make houses smaller in order to make them attractive to the first-time home buyer. Costs can only be cut by reducing floor area.

Trends in envelope costs in houses do not seem to be following trends in office buildings. In Figure 4-14, one can see that the median cost percentage taken by envelope expenses has decreased. This could represent a continuation of the historical decline seen in Figure 3-8, as home owners place more of an emphasis on interior comfort than on exterior decoration. There have also been numerous innovations in envelope materials and methods of construction. For instance, plywood and composition panel siding have become more common on houses. These are not only less expensive than wood clapboards and the like, they require less handling and labor on the

4-14. Recent Cost History of the Envelope—Single-family Residences.

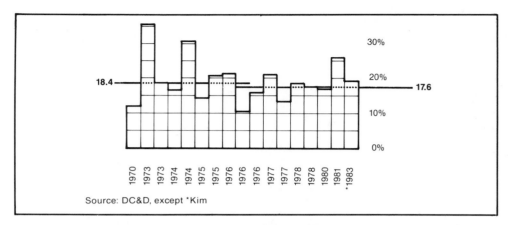

Source: DC&D, except *Kim

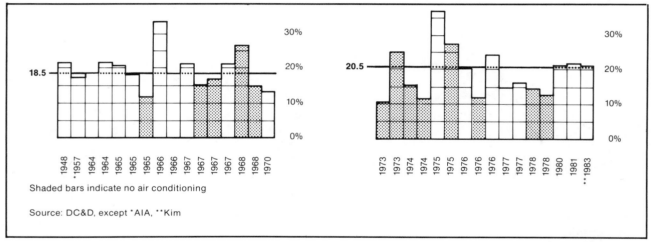

Shaded bars indicate no air conditioning

Source: DC&D, except *AIA, **Kim

4-15. Recent Cost History of the Machinery—Single-family Residences.

site. In addition, new methods of stapling shingles to roofs have eliminated even more labor. Technological advances have not bypassed office-building envelopes, however, so it is surprising to see the decrease for houses as opposed to the increase for office buildings. Perhaps the best explanation is that envelope improvements due to requirements for energy efficiency were concentrated in houses on insulation—a relatively inexpensive feature—whereas office buildings saw improvements in glass and other costlier materials. Additionally, while increased ornament on the large building envelope has recently driven up its cost, most residences were never "unornamented."

While the direction of trends in machinery costs for residences is the same as for office buildings, the degree of change is different. Electronics has made a significant impact on the home. In the past decade or so,

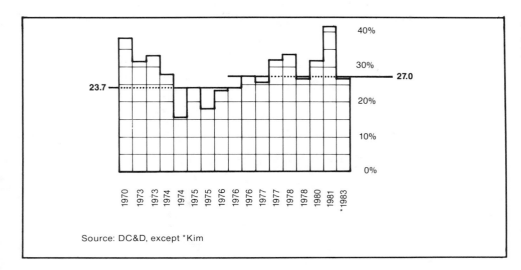

Source: DC&D, except *Kim

4-16. Recent Cost History of the Infill—Single-family Residences.

televisions, stereo systems, microwave ovens, and lighting and temperature controls have gone from luxuries to commonplace household accessories. Other devices, such as ice makers, alarm systems, pneumatic cleaning systems, and trash disposals, bear witness to the increasing introduction of machinery-related devices into the home. In addition, computers are often found in the home now. All of these have generated an increased need for electrical and HVAC advances in the home. Figure 4-15 illustrates that the median cost/percentage for the machinery in residences has gone up in recent years, even though the second half of the sample has more homes without air conditioning than the first half. The increase of two percentage points in the second half of the sample is less than such increases in office buildings, probably because of the sophistication of life-safety devices in larger buildings. Additionally, sprinkler systems and alarm systems are extensive in many offices, but are not commonplace in houses. The machinery's relative cost is likely to continue rising as more labor-saving devices are introduced and as the ones that exist become more commonplace.

In Figure 4-16, a significant rise in infill costs is seen in the median difference between the first half and the latter half of the sample. It is difficult to pinpoint the reason for this and thus to establish whether this trend is or is not likely to continue. The basic features of the residential infill have been the same for over thirty years—partitions, interior finishes, interior architectural decoration, and built-in cabinetwork—and are likely to remain so. What Figure 4-16 may suggest is that the quality of materials used in interiors of houses has increased in recent years. Another reason

4-17. Projection to 2000 A.D. of Component Cost Percentages in Residential Building Types.

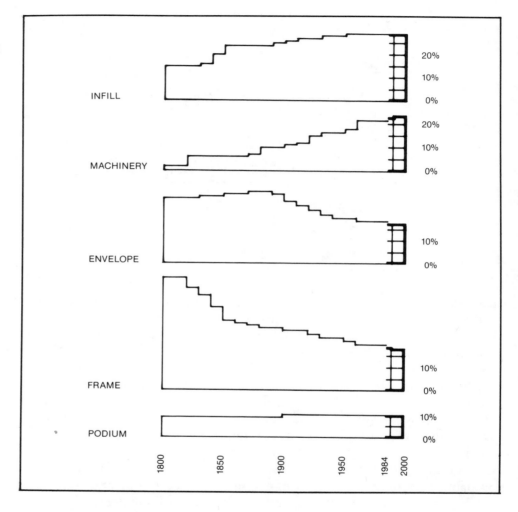

for the rise in the infill's proportion of total building cost is more basic and is similar to the situation in large buildings: the frame cost has declined, and space division has thus been accomplished to some degree by use of the infill.

More than any other component, the infill is a discretionary feature of the building, especially in homes. It is the component that is the least essential and also the component through which the home owner or architect can most express aesthetic preferences. The extent of its use, and hence its cost, may depend somewhat on the tastes of the designer. Since it is the least essential component, the infill is also most likely to be cut back for budgetary reasons. Thus, its cost may depend on the economic climate that prevails at the time of construction. This notion appears to be supported by Figure 4-16, especially in light of the fact that the United States economy was in recession in the mid-1970s—a period that coincided with a drop in relative infill expenditures.

In Figure 4-17, component costs shown in Figure 3-12 are projected to the year 2000 A.D. As with major buildings, residences have been enjoying a period of relative stability since 1950. It is likely that the machinery will rise in relative cost, at the expense of the frame and the envelope. The podium and infill will probably remain at their current levels. These expected changes will not be the herald of major transformations in building technology, however. Rather, they represent the continuing gradual mechanization of the home, a process underway for well over a century now.

CHAPTER 5

The Correlation Between Component Costs and Architectural Design

The Historical Relationship of Cost to Style

Through an examination of the evolution of major stylistic, technological, and economic ages in the history of Western building construction, the interaction of these factors may be seen. The generative influence of construction economics on the development of style becomes quite evident.

In Figure 5-1, the principal architectural styles are summarized. In this context, the Classical style refers principally to Roman Imperial architecture. Its physical characteristics included its massive concrete and "brick" veneer structure, the arch, and the dome. As a style, it was massive, severe, and borrowed heavily from Greek influences, particularly the Ionic, Doric, and Corinthian orders. Major structures of this era were normally used for civic purposes.

Romanesque architecture owed its longevity to the lack of communications and technological advance associated with the Dark Ages. As its name implies, it derived its stylistic features from the fallen Roman Empire. This architecture was massive, constructed in stone, and characterized by the barrel vault and lack of ornament. Most major Romanesque buildings were ecclesiastical in nature.

In the use of the pointed arch, groined vaults, and flying buttresses, Gothic architecture represented the first break from the massive styles that preceded it. Structural innovations in this era significantly opened up the structure to allow large areas of stained-glass windows to appear. As in the Romanesque period, most major Gothic works were church related.

Spurred by socioeconomic advances during the Middle Ages, the Renaissance is marked by the rediscovery of forms from the Classical period

5-1. Evolution of Architectural Styles.

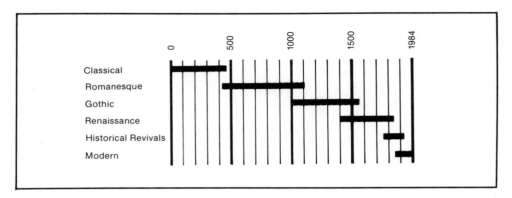

and their application to building design. Renaissance designers did not replicate ancient buildings; rather, Classical motifs were adapted to contemporary structures. Later in the Renaissance, the connection with Classical forms became more tenuous. Thus styles as disparate as the baroque, rococo, Palladian, and Georgian are included in this era. Renaissance buildings—particularly the rococo—exhibited considerably more ornament than their predecessors, especially with regard to infill materials. Principal building types of this age varied from churches to palaces, and from commercial buildings to civic structures.

In the 1700s, the Classical Revival began in England. This ushered in almost two centuries of architecture, especially that of the nineteenth century, that relied on replicating ancient buildings and "shoehorning" modern functions into them. There followed Greek Revival, Egyptian Revival, Gothic Revival, and so forth. H.G. Wells, in his *The Outline of History*, stated that the only style that did not appear in the 1800s was the nineteenth-century style. Major structures of the Revival period were used primarily for commercial, civic, and transportation purposes. This era reached its zenith, and began its rapid decline, at the Columbian Exposition in Chicago in 1893.

Modern architecture as a style began in the late 1800s and has reached its zenith in the post—World War II era. Its signal characteristics have been its expression of steel skeleton or reinforced concrete structures and the use of metal and glass curtain-wall systems. Large Modern buildings are mainly commercial in nature.

Figure 5-2 shows the great technological eras in Western civilization as distinguished by the noted social philosopher Lewis Mumford in his 1931 book, *Technics and Civilization*. Three technological periods were described by Mumford, each characterized by a particular source of power and materials.

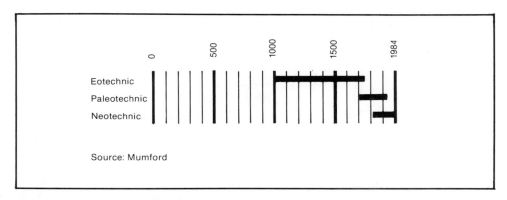

5-2. Evolution of Technological Eras.

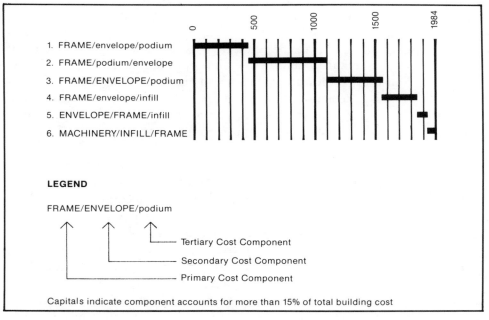

5-3. Evolution of Economic Eras.

The eotechnic phase, stretching from about 1000 A.D. to 1750, used water for power and wood as its material. (Mumford claimed that the impressive masonry structures of this era were dependent on the elaborate wood centering, or formwork.)

The paleotechnic phase, the age of coal and iron, began in 1700, peaked in 1870, and was in rapid decline by 1900. Mumford cites Joseph Paxton's Crystal Palace as this period's crowning achievement. The neotechnic phase began in the 1830s and is continuing to this day. It is characterized by electrical power and alloys.

Figure 5-3, an abstraction of Figures 3-6, 3-7, 3-11, and 3-12 defines the major component cost distribution eras by identifying the components that were the primary, secondary, and tertiary consumers of the building's budget. As noted earlier, more recent times exhibit a shift of the financial resources from the frame to the other components.

Comparisons of the evolution of architectural styles and technologies with the evolution of building costs yield some interesting insights. A significant time lag seems to exist between the introduction of a new technology and its large-scale application to major building projects. This may be seen by contrasting Mumford's eras with the dates at which technologies made an economic impact on the cost of building construction. For example, iron may have been an industrial staple as early as 1700, but its affect on building cost was not evident until after 1850. Electric light was known in 1857 but not applied until the late 1800s. Perhaps the principal reason for this gap is a conservative market. Building owners are investing large sums of money and are reluctant to take risks—they want buildings that they know will work. Thus, innovations have historically been tried in small structures first. This lapse has decreased in recent years, however, due to better communication—via such media as technical journals and advertising—and modern engineering design and testing methods that predict product performance more accurately.

A further chronological gap appears between the date a technological advance affects the economic constitution of a building and the time when stylistic characteristics are incorporated that more accurately reflect this constitution. For instance, the Crystal Palace showed the world in 1851 the technological and aesthetic possibilities of mass-produced prefabricated glass and steel construction. It was nearly a century, however, before such a construction method was fully exploited by International Style designers.

Historically, there have been two factors that contributed to this gap, the first being resistance of building owners to "nonstandard" tastes. Second, architects' professional reputations and hence marketability have always depended to some degree on acceptance by their peers. Thus, conformance to accepted aesthetic norms is professionally profitable. Designers will use new materials to imitate old ones for a time, before learning to exploit them in new ways.

Each stylistic era can be as easily defined by its economic characteristics as by any stylistic particulars. The Romanesque was the era of the frame. The Gothic era was the first in which economic resources were devoted to any great extent to the envelope. In the Renaissance period, money was spent on the infill. The Revival era was characterized by a lavish, albeit short-lived, expense on envelope materials. Technology and style, as well as economics, were in radical transformation during the Revival era, and it is as accurate to view this era as a transition between the Renaissance and modern times as it is to define it as a period unto itself. In fact, this period may be the only one in which style drove cost to any extent—resulting in the brief economic time period shown as era 5 on Figure 5-3.

Percentages of cost allocated to the different components of buildings give evidence of a modern period beginning in about 1950 and continuing to the present. By this time the impact of industrialization is clearly recognizable. The impact of sheet metals and plate glass is apparent in the relatively low envelope cost. Electricity and HVAC, the principal constituents of the machinery, are the signal characteristics of this modern period. Ironically, the machinery is barely visible in most areas of buildings, despite its substantial cost and spatial requirements.

Economics and technology have historically been intertwined in a dynamic and synergistic relationship that largely determines the form of buildings (Figure 5-4). History abounds with examples of innovations that came about through efforts to find a less expensive substitute for a particular building material. Gothic builders realized that glass walls were not as costly as masonry ones, much as Joseph Paxton did centuries later. Later Renaissance architects saw that paint on plaster could give the same effect as exquisitely cut and finished stone. Modern veneers accomplish much the same effect. In the nineteenth century, man discovered that by electroplating he could make the crudest of materials glitter like gold. Economics has driven technological advance, and only after designers belatedly realize the inevitability

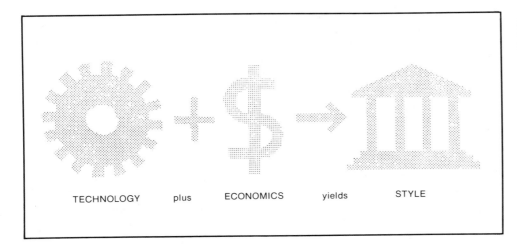

TECHNOLOGY   plus   ECONOMICS   yields   STYLE

5-4. The Interrelationship of Cost, Technology, and Style.

of new building systems and materials do they make a wholehearted effort to work accordingly. Architectural design theory has and must follow the determination of the economic configuration of the building.

A closer look at the period since the onset of industrialization will reveal more closely the recent interaction of technology and cost and how this process gave rise to Modern architecture. Architectural historians have by and large oversimplified the processes of the technoindustrial revolution—the period of the greatest transformation in the economic and technological constitution of buildings. By focusing on the stylistic consequences of the introduction of steel and glass to building, they have overlooked the fact that there were actually several revolutions:

1. Traditionally hand-crafted secondary trades were industrialized (1800–1900). Thus, concurrent with the separation of the frame from the envelope came mass-produced filler materials to replace load-bearing construction.
2. Iron and steel divorced the frame from the envelope (1851–1880). Later, the application of steel to concrete made possible the development of a reinforced concrete as well as a steel skeleton structure.
3. Newly discovered metals and alloys, as well as industrial processes, were applied to the envelope (1900–1950). Examples are aluminum and steel sheets, plate glass, and precast concrete panels.
4. Synthetics, such as sealants, laminates, and resilient flooring were applied to the envelope and infill (1900–1980).
5. Electrical power invaded buildings (1890–1950).
6. Heating, ventilation, and air conditioning (HVAC) were made feasible by the development of electricity and sheet metals (1930–1960).

These six "revolutions" are illustrated in Figure 5-5 on a chronological chart that also plots the evolution of the concurrent economic and stylistic eras. It is easy to see in this figure the direct relationship between technology and cost. The frame dropped to secondary cost importance after it was divorced from its nonstructural purposes. The envelope's brief prominence as a cost component occurred between revolutions. The machinery and infill took over as the most expensive cost elements almost immediately after major innovations occurred in each.

In Figure 5-5 the extent to which the convergence of technological and economic forces affected the course of stylistic development is apparent. Technological developments separated the superstructure into a discrete component—the frame. Since it was also the most expensive component, there was pressure on designers to economize in its construction. The com-

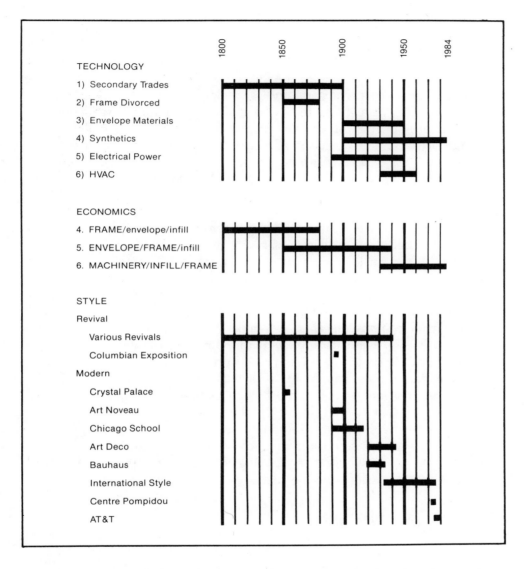

5-5. The Interrelationship of Cost, Technology, and Style in the Industrial Age.

bination of these forces yielded the Chicago School of architecture, where a simplified steel frame was coated with a richly ornamented envelope. Designers such as Louis Sullivan expressed this basic organization of the building's components, rather than disguising it in neoclassical garb. Art Deco in many respects continued this tradition.

After the turn of the century, metals, alloys, glass, synthetics, and the like revolutionized the construction of the envelope. At the same time, the envelope was peaking as an economic component. The desire to cut costs on the big-ticket items—the frame and the envelope—combined with the new technologies to carry the Chicago School's ideas a step further. The Bauhaus in Germany pioneered the "glass and steel" aesthetic, with its use of standardized, repetitive parts and large flat surfaces of sheet metal and plate glass. This gave birth to the so-called International Style that has dominated architecture for much of the past forty years. The overall decrease in weight of the building due to the lightening of the envelope also led to a lighter and less expensive frame.

More recently, the machinery has further revolutionized buildings, and it is now the primary cost component. At the same time, several innovations in the technology of the infill have culminated in its being the secondary cost component. Once again, the desire to save money on the major cost components affects design decisions and generates aesthetic expression. In contrast with the earlier phases of the Industrial Revolution, however, the relative expense of the big-ticket item—the machinery—is likely to continue to rise. The natural result of this condition is that economy will be sought in the construction of the infill, as well as the frame. The open plan has been a partial response to this. The most recent phase of the Industrial Revolution is, however, incomplete at this time, for the ultimate aesthetic expression of economic age 6 has yet to surface.

Design has not yet caught up with technology and economics in the Modern era, principally because most tenets of Modernism were formulated prior to World War II and put into practice in the 1950s building boom. Many of those involved with the famous Bauhaus and International Style did not, and probably could not have foreseen the enormous impact of the machinery on today's buildings. As a result, the disciples of Modernism have always been uncomfortable with this component.

The Direction of Design

It is clear that the era of startling innovation in building technology is largely behind us. The basic makeup of buildings has remained essentially unchanged for over thirty years and is likely to remain so in the foreseeable future. We are in a period not dissimilar to the Decorated Gothic or rococo, when basic building technology was in place and the frontier of design was to explore the manner in which it was adorned.

In the 1970s, the designs of two monumental structures were widely publicized, and construction on both commenced. Both are noted in Figure 5-5, and Centre Pompidou in Paris is illustrated in Figure 5-6. The AT&T

Corporate Headquarters Building in New York has been shown earlier. These two structures illustrate contrasting efforts to adorn present-day buildings.

In the AT&T Building, the decorated envelope that disappeared during the peak of Modern design has returned, now that its cost is not so prohibitive. The AT&T Building as a whole, however, is probably more the signal of a temporary pause in the history of design not unlike Art Noveau. The Postmodern AT&T Building represents the architectural establishment's official discarding of the severe asceticism of the International Style, much as Art Nouveau heralded the rejection of the Revival styles. While Postmodernism and Art Nouveau have acted as cathartics, the underlying march of technology and economics has rendered them both short-lived. The Portland Civic Building shown in Figure 5-7 is another example of Postmodern design, with its multicolored stucco facade surrounding an otherwise typical office interior.

Centre Pompidou in some respects represents the completion of the technoindustrial design revolution. It has demonstrated that the machinery may be a component of visual interest and vitality. It appears that architects have finally taken their cue from the various refineries, power generating plants, and other utilitarian structures that are so prominent on the landscape of the industrialized world. Figure 5-8 illustrates a side elevation of the Arizona Public Service power plant in Tempe. As a sculptural object, the

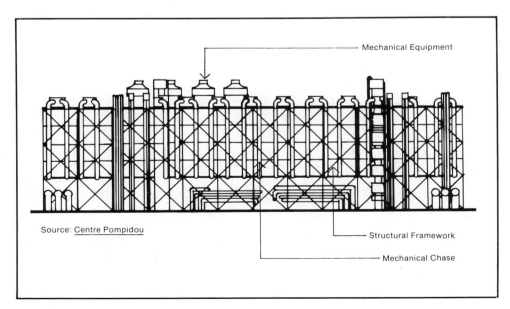

5-6. Centre Pompidou.

5-7. Portland Civic Building.

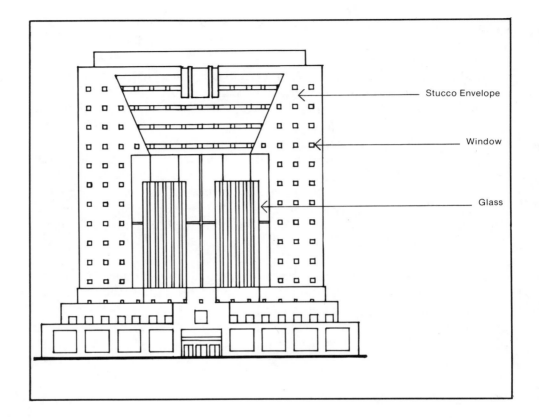

Stucco Envelope

Window

Glass

Arizona plant—with its bilateral symmetry and restrained, less selfconscious use of color—is superior to Centre Pompidou, with its disorderly-looking facades and garish colors.

The necessity of exposing the machinery in order to eliminate money spent on the infill will increase as humankind settles increasingly inhospitable environments—deserts, outer space, other planets—where environmental control is essential for survival.

If there is currently confusion in architectural design, part of the blame lies in the process itself. When the labor of building design divided and the separate design professions proliferated in this century, consensus, rather than vision, became the hallmark of design decision making. This has resulted in an integration of the technical aspects of building based on a "piece of the pie" approach, by which each professional gets a particular economic slice of the design pie. This works to inhibit the development of an integrated approach to the building's aesthetic.

Furthermore, architects have attempted to compensate for their decreased role in the design of the building's components by complicating the geometries of those components over which they still maintain some control. The architectural profession has a vested interest in increasing envelope and infill costs, even at the expense of an integrated building design. Simple economics, however, will doom this effort.

An acute need exists for the architect who is not merely a designer of the envelope and infill, but who is an overseer versed in all trades and technologies. Like automobiles, buildings are complex and modular, and the visible building is the physical manifestation of the integration of essential systems. As Womack and Jones pointed out, the designer must tune these systems to work together harmoniously. Building design cannot afford to overlook the very real aesthetic potential of the 65 percent of the building present in the podium, frame, and particularly the machinery. Designers

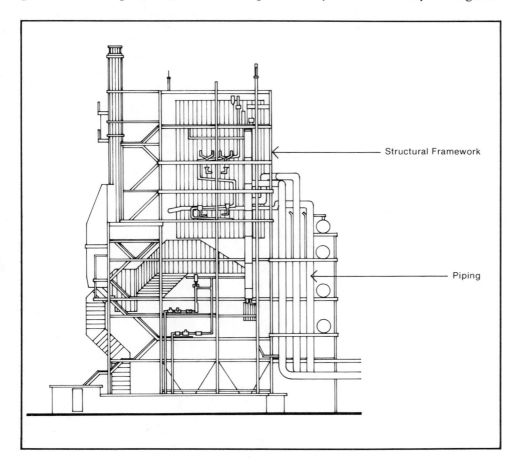

5-8. Arizona Public Service Power Plant.

Structural Framework

Piping

will be wise to look to the design principles evidenced in buildings throughout the centuries—architecture is the organization, arrangement, and ornamentation of the essentials of building.

Modifications to the construction process would be of benefit to the art of building as well. In the Middle Ages, freemasons and sculptors emerged from the ranks of the mason's guild to form the branch of tradesmen that shaped the crude pieces of stone into handsome ornament. Today a similar emergence from the ranks of the electrical and mechanical crafts would be an invaluable asset in the refinement of these parts of the building for decorative purposes.

Great architects throughout history have realized or sensed the economic disposition of the parts of buildings in their times. They disposed of these parts to achieve the finest aesthetic effect. Not until architects again understand the economic basis of the decision-making process will they advance the theory and practice of architecture.

Historical Development of the Podium

Year	Foundations	Paving	Vertical Transportation
0 A.D.	*ca. 100 B.C.* Roman homes use foundations of concrete or masonry rubble brought up to the level of the first floor	*ca. 200 B.C.* Romans use large flat pieces of stone for paving	*ca. 100 B.C.* Mechanical lifting devices have been used in underground mines at least since Ancient Roman times
	ca. 476 Decline of the building arts is a marked feature of the Dark Ages	*ca. 200 B.C.* Etruscans and Romans use paved floors of beaten earth, clay, or stone	
	ca. 900 Wall foundations taken to bedrock at Lincoln, England	*ca. 900* Saxons make neither brick nor tiles	
1000 A.D.	*ca. 1000* Concrete still used for foundations and cores of massive piers in Middle Ages		
	ca. 1100 Foundations made by trenches, somewhat wider than the wall above, filled with rubble stone bound with mortar	*ca. 1300* Rubble stone used for paving	
		1390 Marble pavers used at cloister at Exeter	
		ca. 1400 Inlaid paving tiles find greater use	
			1500s Water-driven mine hoists are used
1700 A.D.	*ca. 1750* Concrete revived in foundations when Semple uses hydraulic lime/gravel mix in foundations of Essex Bridge in England		
1800 A.D.	*1824* Portland Cement, first artificial cement, is invented		
	1833 Founder's Hall at Girard College in Philadelphia and U.S. Sub-Treasury in New York use masonry arch foundations—cellars put to use		

Year	Foundations	Paving	Vertical Transportation
1800 A.D.			*1854* Otis demonstrates elevator safety catch
			1857 Otis elevator is installed in New York
	1871 Cement industry established in U.S.		*1870* More than 2000 lifts are in service in USA
	1877 Reinforced concrete patented by Monier		*1878* Hydraulic lift invented
	1886 Ponce de Leon Hotel in Florida is first monolithic reinforced concrete structure		
	1800s Systematic use of concrete for structural purposes begins in U.S. in late 1800s		*1889* Electric lift invented
1900 A.D.			*1903* Gearless traction for elevators invented
	1910 Reinforced concrete becomes widely used in U.S.		*1924* Automatic controls in elevators invented
	1950s Caisson wells technique mechanized, cutting drilling and concreting costs to 25 percent of previous costs		*1950* First elevators without attendants are widely used

Historical Development of the Frame

Year	Concrete	Masonry	Iron and Steel	Wood
		ca. 300 B.C. Some Roman villas use adobe wall construction up to 200 B.C.		*ca. 300 B.C.* Etruscan house at Marzabotta uses timber frame
				200 B.C. Wooden columns and beams utilized by the Romans in house construction up to 200 B.C.
0 A.D.				
	ca. 100 Rome was made chiefly of concrete		*ca. 100* Iron clamps used to hold wythes of bearing walls together	*ca. 100* Romans use wood screws, an invention that disappeared with the collapse of the Empire
	ca. 100 Roman "opus incertum" method of rubble/mortar wall construction with veneer of no structural consequence is in wide use			
	ca. 100 Romans pioneer vaults and domes, expand span of arches			
	ca. 100 Roman building codes emphasize safety rather than economy of materials—thus, massive construction			
	ca. 100 Slave labor and strict penalties for inadequate structure result in "no great inducement to structural economy in Rome"			
	123 Pantheon completed with span-thickness ratio of 11:1			
	123 Pantheon built of concrete faced with brick veneer as permanent formwork			
	476 Fall of Rome signals collapse of slave economy, ushering in age of labor shortages	*410* Buildings are small, with thick walls, as a result of insecurity in late Rome		
500 A.D.		*ca. 500* New masonry buildings modest in size and primitive in construction		

Year	Concrete	Masonry	Iron and Steel	Wood
500 A.D.		*ca. 700* Thick walls essential for stability of Romanesque vault—it is dangerous to pierce them with large openings		
		ca. 900 Sculpture in Romanesque buildings is part of structure of building		
1000 A.D.		*1000* End of era of small buildings with thick walls		
		1000s Romanesque architecture develops most vigorously—outward thrust of vault offset by cross-vault or thick walls		
		1100 Gothic style begins 400 years of dominance in Western Europe	*ca. 1100* Iron clamps used in masonry construction of medieval period	
		1137 St. Denis Abbey church is first burgeoning of Gothic architecture	*1100s* Architectural detail becomes more elaborate	
		1160 Triforium passage represents major step in the skeletonization of the structure	*1100s* Face stones of thin walls in Gothic buildings carry loads; rubble core acts as bond between wythes	
		1100s Groined vault in Gothic construction concentrates load at corners, liberating walls from their functions as supports		
		1100s Gothic structural system invented		
			1245 Labor shortage occurs due to Black Death	*ca. 1200* Gothic masonry construction not popular for domestic buildings
			1200s Labor and money shortages result in economic use of stone in cathedrals	
		1400 Gothic is universal style of building in the Western world		*1400s* English timber-framed houses are common
		1412 Ornamental and rough stone cost same at King's Hall in Cambridge		
		1400s Early in century, stock carved stone pieces are available from quarries		
		1434 S. Maria del Fiore built in Florence with span-thickness ratio of 21:1		
		1450 Renaissance styles sweep through Europe	*mid-1400s* Threaded nuts and bolts appear	*mid-1400s* Threaded nuts and bolts appear

Year	Concrete	Masonry	Iron and Steel	Wood
		ca. 1450 Renaissance utilizes brick and stone as structural materials		
1500 A.D.			*1560* Iron tension rings used in St. Peter's dome	
		1622 Bricks commercially made in New World		*1600s* American Pilgrims "invent" wood stud wall
				1631 First power saw built in New World or England, in Maine
		1710 Structural brick dome supporting false inner and outer shells is built at St. Paul's in London with span-thickness ratio of 36:1		*mid-1600s* "Saltbox" wood framed house used by New Englanders
		1750 Renaissance styles have completely ousted Gothic in even the remotest parts of Europe		
		1750 Industrial Revolution marks greater use of brick		
			1779 Cast iron begins to be substituted as structural material for masonry	*late 1700s* Steam-powered saw mill enables mass production of standardized lumber
			1792 Strutt uses "fire-proofing" on timber by use of plaster	
			1800s Perfection of extensometers for measure of small strains puts elastic theory on sound experimental basis	
1800 A.D.			*1803* Strutt builds iron frame mill	
			1800–1810 Maudslay's lathe modernizes screw making	
	1822 Frost's Cement patented in England	*1820* Reinforced masonry appears in England	*1822* Iron introduced into American building as auxiliary material used in essentially masonry buildings	
	1824 First artificial cement (Portland Cement) patented in England		*1830s* Inexpensive machine-made nails make balloon frame possible	*1830s* Nails enable development of the balloon frame
				1840 Typical U.S. house built completely of milled lumber
	1852 American Society of Civil Engineers established		*1851* Bogardus's cast iron buildings constructed in New York	
	1854 First reinforced concrete patent by Wilkinson		*1851* Crystal Palace built in London—first major prefabricated and site-assembled iron structure	

Year	Concrete	Masonry	Iron and Steel	Wood
1800 A.D.			*1860* Bessemer patents tilting converter	
	1868 Prefabricated concrete blocks begin to be made commercially		*1870* Total world steel output is 500,000 tons	
	1877 Monier patents reinforced concrete beams			
	1879 Hennabique builds reinforced concrete floor slabs		*1881* Produce Exchange in New York is first building to carry all loads on iron skeleton	
	1885 Koenen advances first theory of reinforced concrete design		*1885* Rolled steel joists produced for first time	
	1886 Ponce de Leon Hotel in Florida is pioneer work in monolithic concrete structure		*1885* First building conceived as iron and steel skeleton frame—Home Insurance Building in Chicago	
	1888 Ransome builds reinforced concrete buildings		*1890* First complete steel frame building constructed in Chicago	
			1899 Benjamin Talbot builds first furnaces for open-hearth continuous process	
1900 A.D.	*1900* Maillart patents flat slab reinforced concrete			
	1908 Albert Kahn work establishes form of standard concrete framing—later variations simply improve mix or reinforcing distribution			
	1927 Planetarium at Jena, Germany is first reinforced concrete dome designed by the membrane theory, with span-thickness ratio of 420:1			
		1930s Significant use and rapid growth of reinforced masonry after 1933 Long Beach earthquake	*1930* Arc welding adopted for industrial buildings	
	1950s Concrete industry rapidly expands in post-war era		*1950* Welding adopted for high-rise construction	
	1950s Concrete shells designed with span-thickness ratios of 1250:1 to 1700:1			
			1970s Application of aluminum coating to fasteners increases the ability of fasteners to withstand extreme conditions	

Historical Development of the Envelope

Year	Masonry	Glass	Metals	Other
	ca. 100 B.C. Terra cotta tiles used for roofing in Rome		*ca. 100 B.C.* Iron gratings used on ground-level windows in Rome	
	ca. 100 B.C. Terra cotta, stone, or marble blocks, pierced with openings, used in Rome			
	ca. 100 Romans use marble to dress and finish buildings			
0 A.D.		*ca. 0 A.D.* Window glass is not common in Rome; Vitruvius does not mention it		*ca. 0 A.D.* Decorated wooden door frames used in Rome
		ca. 0 A.D. Romans use glass in homes of the wealthy		
		ca. 100 Small window panes common in Imperial villas	*123* Copper used extensively in Rome; Pantheon covered with gold-plated copper tiles	*100* Insulation first used in familiar ways—slabs of cork used in roofs
		700 Window glass reintroduced; it is another 1000 years before it is commonplace in houses		
		ca. 700 Thick walls with few and small penetrations are essential for stability of Romanesque vault		
1000 A.D.		*1100* English church windows are still commonly unglazed	*1100s* Roofs formed of wood trusses bearing skin of lead	*1100s* Roofs formed of wood trusses bearing skin of shingles or thatch
		1125–50 Church window glazing becomes general, including pictorial glass		
		1160 Stained glass replaces many masonry elements		
		1170–80 Manufacture of colored glass is common		
		1100s Stained glass used in churches in Mediterranean lands		*1200s* Timber-frame roofs are common, whether with a false ceiling, stone vaults, or exposed roof
		1280–1380 Peak of stained glass era		*1212* Use of tile on roofs in London is compulsory

Year	Masonry	Glass	Metals	Other
1000 A.D.		*1300s* Crown glass is common in palace windows	*1300s* Iron locks commonplace—much iron door furniture is ornate	
			1300s Timber frame roofs covered with lead are common	
	1400s Ornamental relief tiles find greater use on walls	*1400s* Glass is in general use	*1400s* Water-driven rollers and rotary cutters produce sheet metal in Europe	
1500 A.D.		*1500s* Greater security in Renaissance era permits large windows		*mid-1600s* Clapboard exterior siding introduced in New England "saltbox" houses
		1500 Walls of Gothic cathedrals are mere frames for the beauties of stained glass		
		1500 Stained glass era ends		
		1617 Inigo Jones pioneers buildings with large glass windows in England		
	1622 Bricks commercially produced	*1685* Crown glass sliding sash windows used by Jones in Whitehall, England		
1700 A.D.				
		1750 Glass polishing is mechanized		
	1769 Coade Stone (terra cotta) work is prominent in London	*1773* Cast plate glass made in England		
		1776 Glass company at Ravenhead, England manufactures cast plate glass up to 160 inches by 80 inches, an increase of 250 percent over blown plate glass		
1800 A.D.			*1800–1810* Maudslay's lathe modernizes screw making	
	1832 First concrete blocks are advertised as "Artificial Stone"	*1832* Lucas Chance uses French process in manufacture of cylinder glass		
	1835 Terra cotta has lost fashion in England; Coade Stone is out of business			*1840* Mineral wool first produced in Wales—full century before it becomes popular as a building insulation
	1800s Walls continue to be built of solid masonry	*1851* Crystal Palace in London—Paxton makes first large-scale, prefabricated, site-assembled envelope, completely divorced from load-bearing functions	*1850s–60s* Bogardus and Badger use ornamental cast iron columns and spandrels on building facades in New York	
	1880s Sullivan uses terra cotta in Chicago		*1884* First application of aluminum made in architecture, to Washington Monument	
	late 1800s Style is to place any desired form of ornament on any economically appropriate structure	*1884* Chance brothers develop rolled-plate process for glass manufacture in England	*1886* Heroult and Hall develop commercial process for separation of aluminum from ore	*1891* Cabot's Quilt insulation batt invented
1900 A.D.	*1903* Grand Central Station: structural steel is covered with costly materials—granite, marble, and limestone	*1904* Fourcault and Libby-Owens develop process for drawing sheet glass directly from molten glass	*1900s* Stainless steel invented early in century	

Year	Masonry	Glass	Metals	Other
1900 A.D.		1905 First safety glass made by gluing sheets of glass to celluloid		1905 Plywood patented by Hetzer
	1912 Canzelman patents precast concrete building system, including exterior wall panels			1920 Insulation board made
		1925 Ford Motor Co. and Pilkington Bros. develop continuous strip method to make plate glass		1923 Phenolic resins enable mineral, and later glass, fibers to be bound into batts
				1931 Neoprene developed by DuPont
				1935 Owens-Corning formed to make and sell fiberglass wool insulation batts
				1938 First glass-fiber reinforced polyester (fiberglass) products
	1940 Terra cotta industry is almost extinct			1940 Rock wool made by U.S. Gypsum
				1943 Dow Chemical Corporation produces silicone; General Electric follows—applications include adhesives, sealants, and coatings
	1945 and after Precast concrete gains widespread favor in Europe		1945 Weidlinger patents steel/aluminum tube space frame with split-ring fasteners	1945 Dow Chemical produces styrene foam
				1945 Polyethylene developed; later used for vapor barriers
				late 1940s Particle board is available
				1946 Fiberglass strengthened by addition of epoxy resins
			1947 Fuller patents geodesic dome	1950s Foamed plastics developed for insulation
		1950 Polyvinyl acetate replaces celluloid in safety glass	1950 Aluminum used on U.N. Secretariat building trim	1950s Sealed curtain wall developed
		1952 Pilkington Bros. invent float glass process	1953 25 percent of all window walls made with aluminum, versus 5 percent in 1949	1954 First synthetic-resin plasters made
		1950s Tempered glass invented	1956 Aluminum production ten times greater than in 1939	1955 Flexible urethane foams introduced
	1959 Kip's Bay Plaza by I.M. Pei pioneers concrete envelope panels—later panels not load-bearing			1960s Building industry takes to plastics
			1962 Cor-ten steel used to clad John Deere Co. building in Illinois	1962 Kynar introduced
				1964 First thermal exterior insulation systems on high-rise buildings
		1970s Multipane and reflective-coated glass is popular		1970s Composition board utilized for exterior siding
				1970s Insulative sheathing panels are common
				1970s Vinyl siding is popular
	1978 Johnson/Burgee reintroduce complex masonry cladding at AT&T building			

Historical Development of the Machinery

Year	Plumbing/Sanitation	Electric/Gas	HVAC
		2500 B.C. Artificial lighting provided by oil lamps	*2000 B.C.* Fixed fireplaces in use
	ca. 300 B.C. Almost every Campanian domus of distinction incorporates a private bath		*ca. 300 B.C.* Greeks and Romans used decorated bronze charcoal braziers for heating
	ca. 300 B.C. Gutters, downspouts, and cisterns used to collect water in early Rome		
	312 to 226 B.C. Lavish water supply brought into private houses by standardized lead pipes		
	ca. 50 B.C. Cisterns are obsolete		*100 B.C.* Roman hypocausts use underfloor ducted air for heating, especially in the Trans-Alpine
0 A.D.	*ca. 0 A.D.* Ten aqueducts supply Rome with 220 million gallons of water daily	*ca. 0 A.D.* Clay lamps burning oil provide basic illumination in Roman homes	*ca. 0 A.D.* Roman insulae (apartments) have no arrangements for heating
	400s Final deterioration of Roman aqueducts		
	400–1800 Medieval standards of hygiene fall far below those of Ancient Rome		
1000 A.D.	*1100s* Diagrams made of water supply systems at Christ Church, Canterbury		
	1237 London receives its first water supply		*1200s* Flues and chimneys first introduced in castles; fireplace adapted to multistory buildings
	1285 Improvements in London water supply completed; beginning of steady improvement		
	1300s Lead in demand for gutters, spouts, and pipes		
	1300s Conduit for draining refuse from kitchens into pit in garden is common in England		

Year	Plumbing/Sanitation	Electric/Gas	HVAC
1500 A.D.	*1500s* Ball-and-chain mine pumps are water-driven		*1500s* Large braziers become fully enclosed metal stoves in Northern Europe; ceramic stoves appear in Eastern Europe at same time
	1660 Water supply pipe system designed for Versailles Palace		*1500s* Mines are ventilated by water-driven fans
	1700s Open-pit privies universal in American country and towns		*1617* Coal era means more chimneys; Inigo Jones's buildings in England are illustrative
	1778 Joseph Bramah patents the water closet		*1744* Franklin's cast iron stove detached from chimney
	late 1700s Piped water supply to houses becomes common	*1786* Galvani discovers electric current	*1784* Watt uses steam to heat his office
1800 A.D.	*early 1800s* Availability and proliferation of cast iron water pipes	*1799* Lebon patents gas lighting	*1800s* Cast iron hot water or steam radiators become prominent
	1808 Hot water heating appears	*1807* First successful installation of gas lighting occurs in cotton mill in England; 900 lights are installed with piped gas	
	1800s Automatic water extinguishing systems in use	*1812* Gas Light and Coke Company receives charter in England	
		1817 Candlelight is still common in major U.S. buildings	
	1834 First modern aqueduct—Old Croton in New York—is built	*1834* Rotating coil electrical generators made commercially	*1840* Central steam heating installed in "The Tombs" prison in New York
	1843 Hamburg, Germany is first important municipality to construct a sanitary sewer system		*1844* Hood's *Warming Buildings* and Reid's *Theory and Practice of Moving Air* are published
	pre–1860 Sizable U.S. cities begin to construct water supply systems	*1857* Holmes successfully demonstrates electric arc light	*1860* Most buildings heated by steam or hot water
		1860 Fixed, semiautomatic lighting systems, fueled by gas, appear in U.S.	
	1870s Baths with running water are common	*1865* Gas light reaches peak, lasts until circa 1900	*1867* Dr. Hayward's "Octagon" house explores convection-ducted ventilation techniques
	1870s Sizable U.S. cities build storm and sanitary sewers	*1870* Average domestic consumption of electricity is about 30,000 candle-power-hours per year	
	1874 Parmalee invents fusible link automatic sprinkler	*1878* Swan and Edison perfect light bulb	
		1879 Edison introduces electric carbon filament lamp with screw socket	
		1879 Gas appliances are in wide use in England	
	1880 American Society of Mechanical Engineers formed	*1880s* Construction of gas supply network commences in U.S.	*1880* American Society of Mechanical Engineers formed
		1882 Edison erects the first electrical power station, in New York	

Year	Plumbing/Sanitation	Electric/Gas	HVAC
1800 A.D.		*1882* Domestication of electricity by Swan and Edison	
		1883 Swan and Edison form United Electric Light Company	
		1884 American Institute of Electrical Engineers founded	
		1885 Incandescent gas mantle is in use	
	1891 Apartment houses with complete plumbing are constructed in Chicago	*1889* Stokesay Court, in Shropshire, England, uses exposed electric light bulbs	
		1897 First telephone exchange in U.S. is made	
1900 A.D.	*1900* Most U.S. city residents have access to running water	*1900* Electric lighting is an accepted feature of urban life; fuse is in general use	*1900* Fireplaces and chimneys by now serve no role in heating
		1900 Main features of modern electrical supply industry are in place	*1900* Gas heating challenges open fires as main source of heat
		1901 Vacuum cleaner invented in London by H.C. Booth	
		1901 Mercury vapor light pioneered by Cooper-Hewitt	*1903* Royal Victoria Hospital, in Belfast, has ducted air
		1907 8 percent of American homes wired for electricity	*1905* Haldane establishes that temperature, humidity, and air movement are the criteria for comfort
		1909 Bakelite made for sale, used as electrical insulator	
		1910 Electrical vacuum cleaner invented	*1906* Cramer uses sprayed chilled water to clean and cool air—calls method air conditioning
	1914 Central boilers are common in houses; they provide hot water supply	*1915* Average domestic consumption of electricity is about 200,000 candle-power-hours per year	*1906* Carrier controls humidity in air
	1921–23 Expansion of enameled sanitary fixtures	*1920* 12 percent of British households wired for electricity	*1906* Carrier patents dew-point control
	1920s Gas-fired boilers begin to appear	*1920s* Black cast iron stoves give way to enameled models	*1906* Nichrome alloy patented by A.L. Marsh is first satisfactory electrical heating coil
		1923 20,000 refrigerators in use in American homes	*1920s* Central heating becomes common in modest U.S. homes
		1930 Electric lighting has superseded gas	*1920s* Thermostats introduced
		1930 Gas is most popular cooking fuel in U.S.	*1922* Carrier's is first fully air-conditioned building, Graumanns's Metropolitan Theater in Los Angeles
		1930s Electric range, dishwasher, and refrigerator come into use	*1926* Bruno Taut's designs use bright colors to emphasize mechanical equipment
		1934 33 million telephones in use in the world	*1928* Milam Building in San Antonio is first fully air-conditioned office building
			1937 PSFS Building in Philadelphia is fully air-conditioned

Year	Plumbing/Sanitation	Electric/Gas	HVAC
1900 A.D.		*1938* First fluorescent tubes containing mercury vapor are made commercially in U.S. G.E. and Westinghouse make "Lumiline" tubes	*1938* Reduced heat output of fluorescent tubes versus incandescent bulbs makes large-scale air conditioning, especially in offices, economically viable
			1939 Carrier supplies central station air-conditioning system and central and centrifugal refrigerating system to cool spray water in Havana cotton mill, opening market for such systems
		1941 80 percent of American homes wired for electricity	
		1950 First commercial electronic computer	*1950* Carrier's Conduit Weathermaster installed in United Nations buildings
		1950 86 percent of British households wired for electricity	*1950* Domestic packaged air conditioning comes into use
		1950s Plastic electrical accessories become plentiful	*1950s* Air conditioning becomes commonplace
		Late 1950s Transistors replace thermionic valves in computers	*1953* Bruno Eck invents tangential fan, making air-handling units possible
		1960 96 percent of British households wired for electricity	*1960s* Domestic central air conditioning comes into use
			1964 Harris County, Texas domed stadium uses exposed ductwork ornamentally
		1976 380 million telephones in use in the world	*1976* Centre Pompidou in Paris exposes mechanical equipment in ornamental fashion

Historical Development of the Infill

Year	Ceilings and Floors	Finishes	Partitions and Doors	Ornamental Work and Casework
		18,000 B.C. Painting and color in use		
		3000 B.C. Plaster used as finish material		
	ca. 100 B.C. Floor coverings in Roman residences are probably only for the wealthy	*ca. 100 B.C.* Fine wall paintings are common in homes of affluent Romans	*ca. 100 B.C.* Romans use unthreaded bolts as door pivots	*ca. 100 B.C.* Wall cupboards and wardrobes nonexistent, but bedroom chests and wall shelving are used in Rome
0 A.D.		*ca. 0 A.D.* Romans use marble for wall decoration	*ca. 100* Vitruvius belittles wattle-and-daub use for partitions	*ca. 100 B.C.* Roman insulae (apartments) have few built-in amenities
1000 A.D.		*1000s* Blyth uses roughly dressed stone covered with plaster		
		1114 Merton Priory is lavishly painted with scenes—a common practice in churches from the early 12th to the 16th centuries		
			Middle Ages Wattle-and-daub common for interior partitions	*Middle Ages* Customary practice is to seal, or panel, rooms
			1295 Boards used for partitions at Cambridge Castle	
		1300s Whitewash used liberally on interior plastered walls	*1327* Partitions in the King of England's wine cellars made of wattle-and-daub ("stud-and mud") construction, coated with plaster and whitewashed	
			1402 Laths replace wattles	
1500 A.D.				*ca. 1500* Italian Renaissance artists and northern European timber shortage spur development of decorative plaster, replacing wood
				1501 Henry VII gives charter to the Worshipful Company of Plaisterers of London

Year	Ceilings and Floors	Finishes	Partitions and Doors	Ornamental Work and Casework
1500 A.D.				*1532* Westminster Palace has paneling and drapery for walls, cornices and battens for ceiling
			1600 American Pilgrims adapt medieval English wood skeleton frame, with brick nogging infill, to colonies—nogging is covered with clapboard on exterior, plaster on interior— eventually the brick is omitted, thus the wood stud wall	
		1700s Only new materials introduced to American building are paint and lime plaster		*1700s* Baroque era revives ornamental plaster for false-vaulted ceilings and coating on walls to hide natural stone— this is painted to resemble natural stone
		1792 Strutt "fireproofs" timber beams with plaster and tile enclosure	*1790* Introduction of nail and spike cutting machines in U.S.	
			1793 Bentham patents woodworking machines	
1800 A.D.			*1814* Power-driven circular saw appears in U.S.	*1816* Woodworking machinery is developed
			1830 Machine-made nail appears	*1830–50* "Arsenal" of substitute materials created; machine-made ornament peaks
			1832 First concrete blocks made	
			1833 Founder's Hall at Girard College in Philadelphia uses masonry domes and vaults, opening up large interior spaces; partitions reduced to non-load-bearing screens	
			1837 Concrete blocks introduced in U.S. by G.A. Ward	*1837* Jacobi invents electroplating; inexpensive metallic coating applied to cheap materials
		1841 C. and J. Potter print wallpaper by machine in continuous lengths	*1840* Milled lumber in standard sizes is common	
			1840s Door locks made of interchangeable parts	
	1860s First synthetic resins invented		*1861* Linus Yale invents pin-tumbler cylinder lock; forms company with Towne	
	1866 Linoleum flooring invented	*1866* Hyatt brothers and Parkes develop first thermoplastics	*1868* Manufacture of precast concrete block on commercial scale inaugurated by Frear in Chicago	
		1882 Start of modern lacquer and varnish industry	*1898* Gypsum partition tile developed	
		late 1800s Method discovered to control gypsum setting time	*1898* Sackett-board, the first plasterboard, is invented—multiple layers of paper are pulled through plaster	

Year	Ceilings and Floors	Finishes	Partitions and Doors	Ornamental Work and Casework
1900 A.D.			*1900* Several manufacturers produce gypsum lath and wallboard	*Early 1900s* Stainless steel invented
			1905 Plywood patented by Hetzer	*1905* Plywood patented by Hetzer
	1906 Kuhn and Loeb Bank in New York uses suspended ceiling to conceal mechanical equipment	*1912* Phenolic resin coating marks revolution in paint and varnish industry	*1915* Adamant Board, a two-paper gypsum lath, is developed	
		1918 Phenolic laminated sheet patented by Westinghouse	*1917* Sackett-board discontinued; sheetrock panels introduced	*1916* Bakelite invented, used for molds
				1918 Phenolic laminated sheet patented by Westinghouse
		1920s Oil-modified alkyd paints begin to replace oil paints		*1920s* Phenolic-resin bonded plywood is developed
	1926 Asphalt floor tile appears			
	1929 Advertising for phenolic laminates includes baseboards, trim, facings, wainscoting, ceiling panels, etc.	*1920s* Urea resins developed; used for bonding of laminates, e.g., Formica	*1927* Rohm makes Plexiglas	
			1930s Foamed core and recessed edges greatly improve sheetrock panels	*1930s* Synthetic resin adhesives developed, opening way for more uses for plywoods and veneers
	1931 Neoprene developed by Dupont			
	1936 Burgess Acousti-Vent is first acoustic tile ceiling	*1938* Decorative laminates get colors, gain popularity	*1936* Rohm and Haas make Plexiglas; Dupont makes Lucite	
		1941 Epoxy resin surface coatings developed		
	1946 Vinyl floor tile is made		*Late 1940s* Particle Board is available	*1946* Epoxy resins combined with reinforced plastics to make fiberglass
	1947 Standardized suspended ceiling systems available		*1950s* Water-repellent face paper and gypsum core are developed	
		1960s Building industry takes to plastics	*1960s* With invention of screwable metal stud, screw gun and self-drilling, self-tapping screw, gypsum board becomes widespread in commercial buildings	
		1965 Paint is cheaper and better than ever		
	1964 Nylon Astroturf pioneered			

Bibliography

Allcott, Arnold. *Plastics Today*. New York: Oxford University Press, 1960.

American Institute of Architects. *Building Cost Manual*. New York: John Wiley & Sons, 1957.

Arthur, William. *Estimating Building Costs*. New York: Scientific Book Corporation, 1928.

Arthur, William. *New Building Estimator's Handbook*. New York: U.P.C. Book Company, 1922.

Banham, Reyner. *The Architecture of the Well-Tempered Environment*. London: The Architectural Press, 1969.

Barnes, Frank E. *Estimating Building Costs*. New York: McGraw-Hill Inc., 1931.

Beard, Geoffrey. *Decorative Plasterwork in Great Britain*. London: Phaidon Press, 1975.

Bedford, John R. *Metalcraft: Theory and Practice*. London: John Murray, 1967.

"Behind a 'Renaissance' Tower: A Contemporary Structure That Lets It Work." *Architectural Record*, October 1980, p. 106.

Branner, Robert. *Gothic Architecture*. New York: George Braziller, 1961.

Brown, Daniel C. "Composite Tower, Steel Base." *Building Design and Construction*, June 1983, p. 1100.

"Case Studies: Offices/Banks." *Design Cost & Data*, 1958–83.

Cassimatis, Peter J. *Economics of the Construction Industry*. New York: The Conference Board, 1969.

Centre Pompidou. New York: Rizzoli International Publications, Inc., 1977.

Chlystyk, Walter. *Painting and Decorating*. New York: McGraw-Hill Inc., 1965.

Christian, Elizabeth. "Elevator Has Come a Long Way Lately Due to Technology." *Houston Chronicle*, 1983.

Condit, Carl W. *American Building*. Chicago: University of Chicago Press, 1982.

Cowan, Henry J. *An Historical Outline of Architectural Science*. New York: Elsevier Press, 1977.

Cowan, Henry J. *The Masterbuilders*. New York: John Wiley & Sons, 1977.

Cowan, Henry J. *Science and Building*. New York: John Wiley & Sons, 1977.

Cowan, Henry J. and Wilson, Forrest. *Structural Systems*. New York: Van Nostrand Reinhold Company, 1981.

Dallavia, Louis. *Estimating General Construction Costs,* 2nd ed. New York: F.W. Dodge Corporation, 1957.

Derry, T. K. and Williams, Trevor I. *A Short History of Technology.* New York: Oxford University Press, 1961.

Dietz, Albert G. H. *Dwelling House Construction.* Cambridge: MIT Press, 1971.

Dietz, Albert G. H. *Plastics for Architects and Builders.* Cambridge: MIT Press, 1969.

Dingman, Charles F. *Building Estimator's Data Book.* New York: McGraw-Hill Inc., 1929.

Dorn, Harold and Mark, Robert. "The Architecture of Christopher Wren." *Scientific American,* July 1981.

"Duarte Town Center." *Design Cost & Data.* 26:11.

DuBois, J. Harry. *Plastics History USA.* Boston: Cahners Books, 1972.

"EMS: $1 Billion and Growing." *Building Design and Construction,* February 1983, p. 82.

Engineering Foundation Conference. *Building Economics: Solving the Owner's Problems of the 80's.* Morgantown: American Association of Cost Engineers, 1981.

Fitch, James Marston. *American Building, The Historical Forces That Shaped It.* New York: Schocken Books, 1966.

Fitchen, John. *The Construction of Gothic Cathedrals.* Oxford: Clarendon Press, 1961.

Giedion, Siegfried. *Mechanization Takes Command.* New York: Oxford University Press, 1948.

Goldberger, Paul. "Terra Cotta Adds Color to City Buildings." *The New York Times,* September 15, 1983, p. 17.

Goldthwaite, Richard A. *The Building of Renaissance Florence.* Baltimore: Johns Hopkins University Press, 1980.

"Granite Panels Perform as Unitized Assemblies and Give Form and Color to IBM's Prism Tower." *Architectural Record,* November 1976.

Graves, Frederick E. "Nuts and Bolts." *Scientific American,* June 1984.

Harrison, Henry S. *Houses.* Chicago: National Association of Realtors, 1973.

Heery, George T. *Time, Cost and Architecture.* New York: McGraw-Hill Inc., 1975.

Herkimer, Herbert M.W. *Cost Manual for Piping and Mechanical Construction.* New York: Chemical Publishing Company, Inc., 1958.

Hicks, Isaac Perry. *The Estimator's Price Book and Pocket Companion.* New York: David Williams Company, 1906.

History of STO. Tucker, Georgia: STO Industries, Inc.

Hodgson, Fred T. *The Builder's Guide and Estimator's Price Book.* New York: The Industrial Publication Co., 1882.

Hodgson, Fred T. *Estimating Frame and Brick Houses,* 2nd ed. New York: David Williams Company, 1902.

"I.M. Pei & Partners." *Architecture and Urbanism,* January 1976.

Ingels, Margaret. *Willis Carrier, Father of Air-Conditioning.* Garden City: Country Life Press, 1952.

Khan, Fazlur R. and Rankine, John, group coordinators. *Tall Building Systems and Concepts.* New York: American Society of Civil Engineers, 1980.

Kihlstedt, Folke T. "The Crystal Palace." *Scientific American,* October 1984.

Klemm, Friedrich. *A History of Western Technology.* Cambridge: MIT Press, 1964.

Knoop, Douglas and Jones, G. P. *The Medieval Mason.* Manchester: Manchester University Press, 1967.

"Learning Resources Center." *Design Cost & Data* 26:11.

Lee, P. William. *Ceramics.* New York: Reinhold Publishing Corp., 1961.

Lipsey, Robert E. and Preston, Doris. *Source Book of Statistics Relating to Construction.* New York: Columbia University Press, 1966.

Lloyd, William B. *Millwork.* Chicago: Cahners Publishing Co., 1966.

Mark, Robert. *Experiments in Gothic Structure.* Cambridge: MIT Press, 1982.

McGuiness, William J. and Stein, Benjamin. *Mechanical and Electrical Equipment for Buildings.* New York: John Wiley & Sons, 1971.

McKay, A.G. *Houses, Villas, and Palaces in the Roman World.* Ithaca: Cornell University Press, 1975.

Means Square Foot Costs 1984. Kingston: Robert Snow Means Company Inc., 1984.

Means Systems Costs 1983. Kingston: Robert Snow Means Company Inc., 1982.

Morris, Richard. *Cathedrals and Abbeys of England and Wales.* New York: W.W. Norton & Co., 1979.

Mumford, Lewis. *Technics and Civilization.* New York: Harcourt Brace and Co., 1934.

Nichols, Edward. *Estimating.* Chicago: American School of Correspondence, 1913.

Palladio, Andrea. *The Four Books of Architecture.* New York: Dover Publications, Inc., 1965.

"Pennzoil Place." *Architectural Record,* November 1976, p. 101.

Pereira, Percival E., chief editor. *Dodge Construction Systems Costs 1982.* New York: McGraw-Hill Inc., 1981.

Petroski, Henry. "The Amazing Crystal Palace." *Technology Review,* July 1983.

Peurifoy, R.L. *Estimating Construction Costs.* New York: McGraw-Hill Inc., 1953.

Peurifoy, R.L. *Estimating Construction Costs,* 2nd ed. New York: McGraw-Hill Inc., 1958.

Pevsner, Nikolaus. *A History of Building Types.* Princeton: Princeton University Press, 1976.

Pevsner, Nikolaus. *An Outline of European Architecture.* Baltimore: Penguin Books, 1943.

"Philip Johnson." *The Architectural Forum,* January/February 1973, p. 101.

Pulver, H.E. *Construction Estimates and Costs.* New York: McGraw-Hill Inc., 1940.

Pursell, Carroll W., Jr., editor. *Technology in America.* Cambridge: MIT Press, 1981.

Pye, David. *The Nature and Art of Workmanship.* New York: Cambridge University Press, 1968.

Quarmby, Arthur. *Plastics and Architecture.* New York: Praeger Publishers, 1974.

Reynolds, Terry S. "Medieval Roots of the Industrial Revolution." *Scientific American,* July 1984.

Roberts, Thomas A. *Accurate Home Estimating.* Milwaukee: Bruce Publishing Co., 1942.

Rose, Thomas. "Cadillac City Hall Anchors Civic Park." *Building Design & Construction,* September 1980.

Roth, Leland. *McKim, Mead and White 1879–1915.* New York: Arno Press, 1977.

"Saint Anthony's Parish." *Design Cost & Data* 25:29.

Salzman, L.F. *Building in England Down to 1540.* Millwood: Klaus Reprint Co., 1979.

Simmons, Gordon. "Workmanship: Key to Good Building." *AIA Journal,* November 1982.

Skeist, Irving. *Plastics in Building.* New York: Reinhold Publishing Corp., 1966.

Stone, P.A. *Building Economy: Design, Production and Organization,* 3rd ed. New York: Pergamon Press, 1983.

Stubbs, Frank Whitworth, Jr. *Estimates and Costs of Construction.* New York: John Wiley & Sons, 1938.

Sullivan, Bill. "Vertical Transport Going Hi-tech." *Building Design Journal,* July 1983, p. 20.

Sweet's Catalog File. New York: McGraw-Hill Inc., 1981.

TEK Reports. McLean, Virginia: National Concrete Masonry Institute.

"The View From the Top Has Too Much 'Roof Garbage.' " *Buildings Design Journal* 1:9.

This Is the Gypsum Association. Evanston: Gypsum Association, 1980.

Underwood, G. *House Construction Costs.* New York: McGraw-Hill Inc., 1950.

United States Department of Commerce. *Statistical Abstracts of the United States 1984.* Washington, D.C.: Bureau of the Census, 1984.

Van Orman, H. *Estimating for Residential Construction.* New York: Van Nostrand Reinhold, 1978.

Vitruvius. *The Ten Books of Architecture.* Translated by Morris H. Morgan. New York: Dover Publications Inc., 1960.

Wass, Alonzo. *Estimating Residential Construction.* Englewood Cliffs: Prentice-Hall Inc., 1980.

Water, Robert A. *The Builder's Guide,* 2nd ed. Washington, DC: Robert A. Waters, 1851.

Weidlinger, Paul. *Aluminum in Modern Architecture.* Louisville: Reynolds Metals Company, 1956.

Wells, H.G. *The Outline of History.* Garden City: Doubleday & Company Inc., 1971.

Williams, Rosalind. "The Other Revolution: Lessons for Business from the Home." *Technology Review,* July 1984.

Williams, Trevor I. *A Short History of Twentieth Century Technology.* New York: Oxford University Press, 1982.

Wilson, J. *Mechanic's and Builder's Price Book.* Buffalo: Thomas and Lathrop's Steam Presses, 1856.

Womack, James and Jones, Daniel. "The Fourth Transformation in Autos." *Technology Review,* October 1984.

Index